PRAISE FOR IT'S OUR TURN TO EAT

"Wrong's gripping, thoughtful book stands as both a tribute to Githongo's courage and a cautionary tale about the dangers of challenging a thoroughly corrupted system . . . A fast-paced political thriller—with echoes of Graham Greene and John Le Carré." —Joshua Hammer, *New York Times Book Review*

"Urgent and important . . . The story Michela Wrong recounts, ranging from the disaster of British colonial rule in Kenya to its disappointing existence as an independent state, focuses above all on corruption, the one evil she, like John Githongo, sees as the source of almost every other African problem . . . *It's Our Turn to Eat* is the third installment of Wrong's epic trilogy of modern Africa." —Benjamin Moser, *Harper's Magazine*

"Kenya is a prime case of the backsliding that has taken place across Africa . . . tribalism, coupled with foreign aid and cold war politics, helped produce staggering levels of sleaze, and eventually bloodshed. Michela Wrong lucidly brings this all together . . . Her book is packed with detail and solid sourcing and tells its story clearly." —Jeffrey Gettleman, *New York Review of Books*

"Hard-hitting and far-reaching . . . Michela Wrong is an exceptionally talented writer . . . This is one of those rare books that delivers more than the title suggests." —*Guardian* (London)

"A lively and detailed account of the looting of Kenya by its politicians . . . A shocking tale told with verve and suspense." —*Times* (London)

"Important and illuminating . . . Wrong's personal involvement scarcely compromises her excellent analysis of Kenya's twin evils: rat[illegible] oints to the fact that corruption and tribalis[illegible]

—[illegible] *orld*

"A compelling book . . . Wrong's [illegible] rt African morality tale . . . Wrong ex[illegible] this story . . . an African Watergate." —*Financial Times*

"A gripping profile of an anti-corruption crusader . . . Githongo . . . is a magnetic protagonist for Wrong's exposé of the machinery of corruption." —*Publishers Weekly*

© PETER CHAPPELL

ABOUT THE AUTHOR

MICHELA WRONG is a distinguished journalist and has worked as a foreign correspondent for Reuters, the BBC, and the *Financial Times.* She writes regularly about Africa for Slate.com and is a frequent commentator on African affairs in the media. Her first book, *In the Footsteps of Mr. Kurtz: Living on the Brink of Disaster in Mobutu's Congo,* won the James Stern Silver Pen Award for Nonfiction. Her second book, *I Didn't Do It for You: How the World Betrayed a Small African Nation,* is a portrait of the African nation of Eritrea. Michela Wrong lives in London.

Also by Michela Wrong

In the Footsteps of Mr. Kurtz:
Living on the Brink of Disaster in Mobutu's Congo

I Didn't Do It for You:
How the World Betrayed a Small African Nation

IT'S OUR TURN TO EAT

THE STORY OF A KENYAN WHISTLE-BLOWER

MICHELA WRONG

HARPER PERENNIAL

NEW YORK • LONDON • TORONTO • SYDNEY • NEW DELHI • AUCKLAND

HARPER PERENNIAL

First published in Great Britain in 2009 by Fourth Estate, an imprint of HarperCollins Publishers.

A hardcover edition of this book was published in 2009 by Harper, an imprint of HarperCollins Publishers.

Photograph on page x by Peter Chappell

FIRST HARPER PERENNIAL EDITION PUBLISHED 2010.

The Library of Congress has catalogued the hardcover edition as follows:

Wrong, Michela, 1961–
It's our turn to eat : the story of a Kenyan whistle-blower / Michela Wrong.—1st U.S. ed.
p. cm.
Includes bibliographical references and index.
ISBN: 978-0-06-134658-3
1. Political corruption—Kenya. 2. Kenya—Officials and employees. 3. Githongo, John, 1965– 4. Kenya—Politics and government—2002–. Title.
JQ2947.A56C69 2009
364.1'323092—dc22
[B] 2008055426

ISBN 978-0-06-134659-0 (pbk.)

HB 04.17.2023

CONTENTS

	Map	viii
	Prologue	xi
1	The Big Man	1
2	An Unexpected Guest	21
3	Starting Afresh	33
4	Mucking out the Augean Stables	41
5	Dazzled by the Light	65
6	Pulling the Serpent's Tail	77
7	The Call of the Tribe	99
8	Breaking the Mould	121
9	The Making of the Sheng Generation	145
10	Everything Depends on the Boss	163
11	Gorging Their Fill	183
12	A Form of Mourning	205
13	In Exile	227
14	Spilling the Beans	247
15	Backlash	255
16	A Plaza Paradise	279
17	It's Not Your Turn	295
	Epilogue	317
	List of Key Characters	336
	Glossary	337
	Acknowledgements	339
	Notes	341
	Index	345

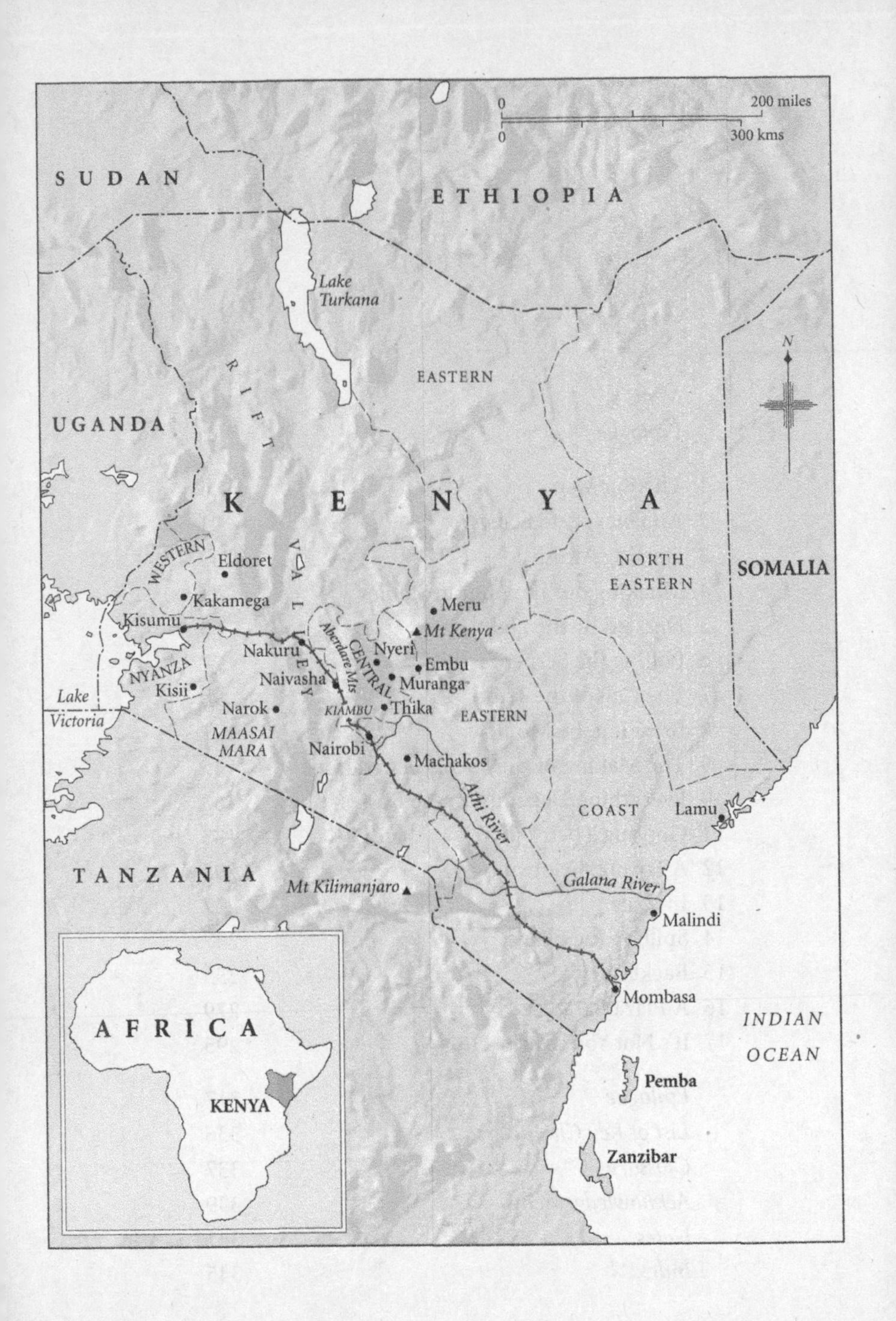
0
200 miles
0
300 kms
SUDAN
ETHIOPIA
Lake Turkana
EASTERN
N
UGANDA
RIFT VALLEY
KENYA
WESTERN
Eldoret
NORTH EASTERN
SOMALIA
Kakamega
Meru
Kisumu
Mt Kenya
Nakuru
Aberdare Mts
CENTRAL
Nyeri
Embu
NYANZA
Naivasha
Muranga
Kisii
Lake Victoria
Narok
KIAMBU
Thika
EASTERN
MAASAI MARA
Nairobi
Machakos
Athi River
COAST
Lamu
TANZANIA
Mt Kilimanjaro
Galana River
Malindi
Mombasa
AFRICA
KENYA
INDIAN OCEAN
Pemba
Zanzibar

'We lived many lives in those whirling campaigns, never sparing ourselves: yet when we achieved and the new world dawned, the old men came out again and took our victory to re-make in the likeness of the former world they knew. Youth could win, but had not learned to keep: and was pitiably weak against age. We stammered that we had worked for a new heaven and a new earth, and they thanked us kindly and made their peace.'

T.E. LAWRENCE, *Seven Pillars of Wisdom*

John Githongo with the author.

PROLOGUE

It was early evening, that time when the traffic jams that clog the Kenyan capital of Nairobi's arteries for most of the working day reach their apogee.

Down below, thousands of honking *matatu* minibuses, drivers hyped on adrenalin and pent-up frustration, were doing their best to get their passengers home. Fighting for space against the lumbering public buses, sagging like old mattresses under the weight of their clientele, the customised Toyotas and Nissans jerked in fits and starts across the crumbling roundabouts, their touts leaning out of half-open doors to wheedle and abuse, pounding on the bonnets of encroaching cars in a manner more bullying than friendly. Choking on the black fumes pumped from hundreds of over-revved, under-serviced engines, blue-uniformed policemen struggled to keep the flow moving.

On the fifteenth floor of the Ministry of Finance, however, the beeps and angry shouts were barely audible. Most of the messengers, clerks and secretaries, keepers of the ubiquitous departmental Thermos of tea, had abandoned their perches behind the varnished wooden partitions, silence was creeping in to take their place. From this elevation, the world below seemed calm and peaceful, lent tranquillity by the approaching chill of evening. Wheeling in cooling eddies of warm air, kites traced monotonous circles high above, like black smuts whirling over a dying fire. Even higher above them gracefully looped Nairobi's sinister valkyries, the marabou storks, hard-working scavengers of the nearby slums.

From up here, the historic landmarks of the city centre, so grubby at street level, looked almost pristine. One side of the ministry gazed south-east, across the rusting, dilapidated entrails of the giant railway depot that was both the city and the country's original *raison d'être*: for Nairobi was the spot where British railroad engineers paused to gather their material, manpower and energies before flinging their ironware up and over the escarpment and dizzily down into the Rift Valley, aiming at the giant lake lying at the continent's alluring heart.

Beyond stretched the hangars and godowns of the Industrial Area, the capital's main airport and the savannah expanse of Nairobi's game park, hemmed in by the dry Ukambani hills, where the odd feather of grey smoke – some peasant clearing land – plumed skyward. The other side of the ministry looked across Harambee Avenue, past the colonnaded Law Courts, towards the clock tower of Parliament Buildings, City Hall and the conference complex named after Jomo Kenyatta, once dubbed a 'leader unto darkness and death' by a British official, now honoured as the nation's founding father. The small dots moving about on the esplanade below were Kenyan sightseers, come to have their pictures taken in front of the late president's seated statue, which showed him in chieftain's cap, flywhisk in hand. Beyond the square one could glimpse the lawns of Uhuru Park, where Mwai Kibaki, Kenyatta's former finance minister, had been inaugurated president eighteen months earlier. From up here, the park seemed the green and pleasant public garden its planners had originally envisaged, rather than what it had become in the intervening years: open-air toilet, haunt of roaming muggers, resting spot for the homeless and exhausted.

Inside the minister's office, three men sat locked in intimate conversation: the finance minister himself, a pudgy septuagenarian with a spray of whitening hair; the justice minister, a former human rights campaigner with an acne-scarred complexion and a woman's pulpy lips; and a third player, a barrel-chested, trunk-necked lumberjack of a man who looked ready to burst from his suit at any moment. What the three were discussing was so engrossing, they were barely aware of their surroundings.

And then it happened. The giant suddenly became aware of a metallic whispering . . . What was that? His stomach lurched as he realised that tinny, tiny sound was coming from his own midriff. He could barely believe it. The recorder he had taped to his stomach, its wire lead and microphone stuck to his breastbone, had somehow switched into 'play' mode. The voices of the two men before him were now being relayed back, potentially exposing him as what he was: spy, sneak, mole.

He coughed loudly, spluttered, coughed again, hoping to drown out the noise. In a booming voice – his voice always boomed, they would find nothing strange in that – he excused himself, lurched out of the ministerial office, and headed swiftly down the gloomy corridor, aiming for the gents'. Inside the cubicle, hands trembling with adrenalin, he adjusted the device. How on earth had that happened? He was tempted for a moment to abandon this particular attempt. Perhaps Fate was telling him not to push his luck any further. But no, might as well be consistent. He had already crossed the line, transgressed in ways that most of his fellow countrymen would have never dreamed possible and many would never forgive. Might as well see the thing through. Routine was important, it lent shape and definition. He put the device back into 'record' mode, carefully adjusted his shirt. He splashed some cold water on his face, took a few deep breaths, and walked back into the office.

He scoured his two colleagues' faces for signs of suspicion. If they had noticed what had happened, he could expect to be arrested that night, his office sealed, staff sent away, files seized, house raided. But the two men hardly looked up. His pounding heartbeat became more sedate. Either they were Kenya's most consummate actors, or they had barely noticed that he had left the room, let alone picked up the whisper emanating from his chest. Cautiously, he resumed his seat. Leaned forward to include himself in the conversation. The recorder was running again. That night he would do what had become a daily chore, summarising the evening's conversation in one of his black Moleskine notebooks, downloading the disc's contents onto his computer, emailing the file in codified form to a friend abroad. Another piece of evidence collected and logged, his insurance against the coming storm. It would all mount up.

Outside, Uhuru Park, Harambee Avenue and the Kenyatta International Conference Centre had all been swallowed up by the darkening sky. Sunset does not last long this close to the equator. The once-busy streets were barely illuminated by the few functioning street lamps, whose dull glow drove the insects crazy but scarcely penetrated the deep African night. They now looked empty and dangerous, delivered over to the city's rapists and thieves. Distant traffic, working its way through the suburbs and outlying slums, gave off a quiet, murmuring rumble. It smelt as though rain was on its way.

1

The Big Man

'It was an amazing thing, for one moment in a hundred years, to all feel the same way. And to feel that it was good.'

Kenyan writer BINYAVANGA WAINANA

A brown clod of earth, trailing tufts of grass like a green scalp, suddenly soared through the air and landed on the stage, thrown by someone high on the surrounding slopes. Then another one sailed overhead, this time falling short and hitting the journalists packed against the podium. Then came some sticks, a hail of small stones. The first rows of the crowd hunched their shoulders and hoped it would get no worse: there were plenty of kids up there from Kibera slum, the sprawl of rusty shacks that stretched like an itchy brown sore across the modern city landscape, and they had a nasty habit of using their own excrement as missiles. The mood in the open-air stadium in Uhuru Park on 30 December 2002, a year and a half before that strange meeting in the finance ministry, was on the brink of turning ugly. Mostly male, mostly young, the audience was getting bored with waiting.

For much of the morning the mood had been cheerful. The thousands of Kenyans who had begun streaming into the amphitheatre at 7 a.m. for the presidential inauguration – the first change of leadership via the ballot box since independence – had every reason to pat themselves on the back. With the simplest of acts, they had pulled off what felt like a miracle. They had queued patiently for hours in the sun, cast their ballots and in the process turned their backs on the retiring Daniel arap Moi, twenty-four years at the helm, the president credited with reducing East Africa's most prosperous economy to '*nchi ya kitu kidogo*': 'land of the "little something"', homeland of the bribe. Campaigning on an issue that infuriated the public – the corruption souring every aspect of their lives – the opposition had united under the banner of the National Rainbow Coalition (NARC) and stomped to victory. It had told the electorate it was '*unbwogable*' – uncrushable – and this had proved no idle boast, for it had broken the ruling KANU party's thirty-nine-year grip on power.

It seemed as though Kenya's political parties had finally matured, realising that so long as they allowed tribal differences to dominate, with each ethnic group mustering behind its own presidential candidate, Moi would win. In contrast with so many of his African counterparts, the loser – Moi's handpicked protégé Uhuru Kenyatta, son of the nation's founding father – had gracefully accepted the results. In the slum estates the night before, many had braced themselves for a military takeover, reasoning that Moi's security services would surely not meekly accept the people's verdict. They had been proved wrong, and the fact that power was about to change hands peacefully in an African nation, rather than at the barrel of a gun, was hailed by the Western press as a tribute to both the rule of law and a politically mature public's self-control. The partying had gone on into the early hours, with Tusker beer washing down roasted chicken. When it became clear which way the vote was going, residents had rounded up all the local cockerels and slaughtered the '*jogoo*', hated symbol of the once-proud KANU, which Moi had promised would rule the country for a hundred years. This morning they were turning up to bear living witness to their own historic handiwork.

Up on the dais, an array of African presidents and generals in gold brocade and ribbons sat fanning themselves. Next to them sweltered the diplomats, ham-pink under their panamas. Kenyan VIPs, finding no seats available, sat uncomplaining on the floor, their wives' glossy wraps trailing in the dust. As the timetable slipped by two, then three, four, five hours, the amphitheatre steadily filled. An incongruous aroma of Sunday lunch wafted through the air as thousands of feet crushed the wild garlic growing on the slopes. Nearby trees sagged under the weight of street boys seeking a bird's eye view. An urchin on the rooftop of the podium wiggled his ragged arse to the music from the military band, which, like all the armed forces present, was beginning to lose its nerve. They had rehearsed exhaustively for this event, but had never anticipated these kind of numbers: 300,000? 500,000? Who could count that sea of brown heads? At the start, police horses had plunged and reared as the General Service Unit (GSU), Kenya's dreaded paramilitary elite, attempted to clear the area in front of the dais. They had pushed the crowd back, only for policemen posted on the fringes to push it forward. But as the throng grew, and grew, and grew, the men from the GSU dismounted and quietly joined the onlookers, aware that the best they could hope for now was avoiding a stampede.

Gathered at the front, we journalists had long ago lost our carefully chosen perches and jealously cherished camera angles, swallowed up by the crowd pressing hard at our backs. Pinned against my neighbours, I could feel small hands, fleeting as lizards, fluttering lightly through my pockets in search of money, mobile, wallet. With a heave, I scrambled onto a creaking table where a dozen sweaty photographers and reporters teetered, bitching fretfully at one another – 'Don't move!' 'Hey, head down, you're blocking my shot!' 'Stop pushing!' – a touch of hysteria – 'STOP PUSHING!' The ceremony was now running six hours late. Rather than whipping up the audience, newly elected MPs were appealing for calm from the stage. A Kenyan reporter next to me rolled the whites of her eyes skywards, gracefully fainted and was passed out over people's heads in the crucifix position, like a fan at a rock concert. I wondered how long it would be before I followed her. People were keeling over left, right and centre, ambulance crews plunging bravely into the throng to remove the wilting bodies.

Finally, amid cheeky cries of 'Speed up! Speed up!', accompanied by 'fast-forward' gestures from the crowd, the ceremony started. An aide walked on bearing a gold-embroidered leather pouffe. This, it turned out, was the Presidential Pouffe, there to prop up the plastered leg of winner Mwai Kibaki, who had survived the years in opposition only to be nearly killed in a campaign car crash. Next came Kibaki himself, his wheelchair carried by eight straining men. The ramp they laboured up had been the topic of a debate which exposed the establishment's nervousness. Frightened of being implicated, at even the most pragmatic level, in this near-inconceivable changing of the guard, jittery officials from the ministry of public works had refused to build the cement slope required, forcing an exasperated army commander to contract the work out to a private firm.

Kibaki was followed by the outgoing Moi, ornate ivory baton clutched in one hand, trademark rosebud in the lapel of a slate-grey suit, face expressionless. Later, it was said the generals had gone to Moi when it became clear which way the election was going and offered to stage a coup. In his prime, his hold on the nation had been so tight, cynics had quipped, '*L'état, c'est Moi.*' But the Old Man had waved the generals wearily away, aware such times were past, Kenya was no longer destined to follow such clichéd African lines.

Eyes yellow and unreadable, Moi took his salute and delivered his last presidential speech without a hint of bitterness, hailing the rival by his side as 'a man of integrity'. This former schoolteacher's presidency had been an exercise in formalism, and he was determined to fulfil this last, painful role impeccably. But the mob showed no mercy – those watching the ends of Africa's dinosaur leaders never do. What fun, after a quarter-century of respectful forelock-tugging, to be able to let rip. 'Bye bye,' they jeered. 'Go away.' Others sang: 'Everything is possible without Moi,' a pastiche of the 'Everything is possible with faith' gospel sung in church. In the crowd, someone brandished a sign: 'KIBAKI IS OUR MOSES'.

Then it was Kibaki's turn. It was a moment for magnanimity – peaceful handovers, as everyone present that day knew, should never be taken for granted in Africa. And the seventy-one-year-old former finance minister, an upper-class sophisticate known for the amount of time he spent on the golf course, his lazy geniality, was not built in the vengeful, rabble-rousing mould. So the concentrated anger of his speech had those sitting behind Kibaki blinking in surprise. It offered a sudden glimpse of something raw and keen: a fury that had silently brewed under the suave façade during years of belittlement. Never deigning to mention the man sitting by his side, his former boss, Kibaki dismissed Moi's legacy as worthless. 'I am inheriting a country that has been badly ravaged by years of misrule and ineptitude,' he told the crowd. He warned future members of his government and public officers that he would respect no 'sacred cows' in his drive to eliminate sleaze. 'The era of "anything goes" is gone forever. Government will no longer be run on the whims of individuals.' Then he pronounced the soundbite that would haunt his time in office, destined to be constantly replayed on Kenyan television and radio, acquiring a different meaning every time. 'Corruption,' he said, 'will now cease to be a way of life in Kenya.' Whenever I hear it today, I notice a tiny detail that passed me by as I stood in that sweaty scrum, smeared notebook in hand, mentally drafting the day's article: Kibaki, always a laboured speaker, slightly fumbles the word 'cease'. Lisped, it comes out sounding very much like 'thief'.

The speeches over, the various presidents headed for their motorcades as the security services heaved sighs of relief. The inauguration had been an organisational débâcle, but tragedy had somehow been skirted, as was the Kenyan way. For Moi, one last indignity was reserved. When his limousine drew away, snubbing a long-delayed State House lunch in favour of the helicopter that would whisk him away from the hostile capital and to his upcountry farm, it was stoned by the crowd.

As I climbed down off the table, my bag momentarily became wedged in the mêlée, and hands reached out from the crowd. Remembering the little fingers at work earlier in the morning, I rounded my shoulders and gave my bag an aggressive yank. 'Oh, no, no, madam,' sorrowed a man, knowing exactly what was in my mind. 'Those days are over now in Kenya, this is a new country.' They were reaching out not to mug me but to help me, a member of the international press who had played a tiny part in Kenya's moment of glory by mere dint of witnessing it. 'You will see, this will be our best ever government,' chimed in a smiling student, sweat-soaked T-shirt plastered to his body, and I felt a spasm of shame.

In the days that followed I would often feel ashamed, for my professional cynicism was out of step with the times. There was a tangible feeling of excitement in the air, a conviction that with this election, Kenyans had brought about a virtually bloodless political, social and psychological rebirth, saving themselves from ruin in the nick of time. Many of those who had represented the country's frustrated conscience – human rights campaigners, lawyers and civic leaders who had risked detention, police beatings and harassment in their bid to drag the country into the twenty-first century – were now in charge. Mass happiness blended with communal relief to forge a sense of national purpose. With this collective elation went an impatience with the old ways of doing things. Newspapers recounted with glee how irate passengers were refusing to allow *matatu* touts to hand over the usual *kitu kidogo* – that ubiquitous 'little something' – to the fat-bellied police manning the roadblocks, lecturing officers that a new era had dawned. There were reports of angry *wananchi* – ordinary folk – storming an upcountry police station to demand refunds of bribes paid over the years. In ministries, at City Hall, at the airport, only the very foolish still asked for the customary backhander. Backs were straightened, desks cleared in nervous anticipation of an incoming deputy minister or mayor out to show the TV cameras that he would have no truck with sloth and incompetence. Large signs – 'This is a corruption-free zone', 'No bribes', 'You have a right to free service' – went up in government offices, along with corruption complaints boxes, which swiftly filled up with letters venting grievances that had festered through the decades.

The social contract taken for granted in so many Western countries, barely discernible in Kenya, suddenly began to make itself felt. 'Damn it all,' a Kenyan writer returning from self-imposed exile told me, with the air of a man making a possibly foolhardy concession, 'I'm even thinking of paying tax!' And just in case anyone was in danger of forgetting the past, the NARC government threw open the basements of Nyayo House, an ugly beige high-rise on the corner of Uhuru Highway, in whose dank cells opponents of the Moi regime had been beaten, reduced to drinking their own urine and killed. When Gallup conducted a poll, it found that Kenyans were the most optimistic people in the world, with 77 per cent saying they had high hopes for the future. Reserved and inhibited, Kenyans are sometimes dubbed 'the Englishmen of Africa' because of their refusal to live up to the stereotype of boisterous, carefree Africans. After decades languishing in the grey fug of the Moi regime, they could barely stop smiling.

And the new president kept hitting the right notes. When the country's biggest companies took out fawning newspaper advertisements congratulating him, Kibaki reproved them for wasting money. He had no intention, he said, of following his predecessor's example by putting his face on the national currency, streets and buildings. His pledge not to bring city traffic grinding to a halt with wailing presidential motorcades seemed to hold good. Across the land, the framed official Moi photograph, so ubiquitous it had become virtually invisible, came down from the walls, but was not immediately replaced with one of Kibaki. Shop owners propped the new official portrait against the walls, waiting to see how the political climate would turn. Perhaps Kenya had got beyond the point of needing such crude symbols of authority. As for the media, they luxuriated in a less fractious relationship with the new establishment. It had taken the dawn of multi-partyism in 1992 for any newspaper cartoonist to dare depict the president. Even then they had gone on tiptoe, initially showing no more than a hand with a *rungu*, Moi's signature baton, then depicting the Great Man as a silhouette from behind, before cautiously shifting him round, image by image, to face the readership. 'It was from simple fear, because they could come for you,' recalled cartoonist Frank Odoi. But with Kibaki, who had been drawn for decades lazing at the golf course, such veneration would have been absurd. The new president was shown full-on, just as he always had been.

The cabinet Kibaki unveiled on the lawn of State House – 'Ladies and gentlemen, we are in business' – rewarded allies who had made victory possible. A member of the Kikuyu,[1] Kenya's largest and most economically successful tribe, Kibaki knew his nation's two-score smaller tribes needed reassurance if they were to stay on board. Announcing the line-up, he promised foreign donors, itching to resume lending frozen during the Moi years of mutual ill-will, that he would swiftly implement two anti-corruption bills dear to their hearts. There was less detail on the new constitution NARC had undertaken to introduce within a hundred days, expected to trim the president's sweeping executive powers and force him to share decision-making with a prime minister. But few doubted this was on its way. A man with a reputation for soft living and hard drinking, Kibaki knew his younger coalition partners had in part rallied behind him because they viewed him as too indolent to want to do much, too old to attempt more than one term. They regarded him as virtually a figurehead, and there was no sign that he intended to renege on the deal.

Things finally seemed to be going right for Kenya, and the news spread beyond the country's borders like a warm glow. 'The victory of the people of Kenya is a victory for all the people of Africa,' South Africa's first lady, Zanele Mbeki, pronounced at Kibaki's swearing-in, and she was right. For Kenya is one of a handful of African nations which have always possessed a significance out of keeping with their size and population, whose twists and turns are monitored by outsiders for clues as to which direction the continent itself is taking. Somehow, what happens here matters more to the world outside than what happens in many larger, richer, more populous African countries.

This pre-eminence can in part be traced to Britain's colonial role and the astonishingly resilient memory of 'a sunny land for shady people', where English aristocrats swapped wives and downed gin-and-tonics while snorting quantities of recreational drugs. Long before Barack Obama's ancestry came to intrigue the Western public, a pith-helmeted fantasy woven from Ernest Hemingway's tales and Martha Gellhorn's writings, the escapades of the Delamere family, stories of the man-eating lions of Tsavo, Karen Blixen's *Out of Africa* and the White Mischief cliché – all references irrelevant to ordinary Kenyans but stubbornly sustained by the tourism industry – guaranteed the country a level of brand recognition other African states could only dream about.

But there are less romantic reasons for Kenya's disproportionately high profile. The most advanced economy in the region – thanks in part to the network of roads, cities, railroads and ports left by the British – Kenya has held linchpin status ever since independence by mere dint of what it is not. It has never been Uganda, where Idi Amin and Milton Obote demonstrated how brutal post-colonial rule could turn; or Rwanda, mourning a genocide that left nearly a million dead; or Sudan, venue for one of the continent's longest civil wars. In place of Ethiopia's feeding stations and Somalia's feuding warlords, it offered safari parks and five-star coastal hotels. Kenya's dysfunctional neighbours have always made it look good in comparison.

It had made the right choice in the Cold War lottery, allying itself with the winning, capitalist side. Kenya was the obvious place to train your soldiers, in the case of the British Army; to moor your warships, in the case of the Pentagon; to base your agencies, in the case of the United Nations; or to set up your Africa bureaux, in the case of Western television and radio stations. The road to the centre of Nairobi from Jomo Kenyatta airport – which services more airlines in an afternoon than many African airports manage in a week – said it all, with its industrial storage depots and hoardings advertising mobile phones and internet servers, beer and mattresses. 'Nairobbery', as expatriates cynically dubbed it, might be potholed and crime-ridden, but it was the capital of a highly cosmopolitan, comparatively stable nation run, through the decades, by a series of administrations Westerners instinctively felt they could do business with. Like its former colonial master, Kenya had always punched above its weight, offering outsiders – wincingly sensitive to the continent's darker manifestations – a version of Africa they could stomach.

So when Kenya, in the latter part of the Moi era, appeared to veer off course, the world pricked up its ears. Moi, admittedly, had been nothing like as crudely predatory as Zaire's Mobutu Sese Seko, Togo's Gnassingbe Eyadema or Cameroon's Paul Biya – contemporaries all. But as diplomats repeatedly told government officials smarting at their criticisms: 'We hold Kenya to higher standards than other countries.' And when measured against what it could have become, rather than against neighbouring basket cases, Kenya, by the turn of the century, was beginning to look desperately unimpressive, the model pupil turning sullen delinquent. The end of the Cold War, which had transformed the prospects of so many African states devastated by the superpowers' proxy wars, had delivered no obvious dividend here. Hopeful talk of an emerging group of 'Renaissance' leaders who would find 'African solutions to African problems' did not include Kenya, weltering in a political, economic and moral miasma. Once ranked a middle-income country, Kenya lagged towards the bottom of the international league tables, its early potential unfulfilled. At independence in 1963, average per capita income had been level with that of Malaysia; now Malaysia's was ten times as high.

Moi liked to be known as the Professor of Politics, and the man dismissed by his enemies as a 'passing cloud' when he succeeded Kenyatta in 1978 had proved a remarkable survivor, riding out a shift to multi-party politics that many had assumed would unseat him. Yet in the process he had pauperised many of his thirty million citizens, of whom 55 per cent now lived on less than a dollar a day. In Nairobi's sprawling slums, the largest and most sordid in Africa, Western-funded non-governmental organisations (NGOs) provided basic services, not the state, of which nothing was expected. When Kenya marked forty years of independence in 2003, newspaper cartoonists could not resist highlighting the cruel trick history had played on the country. They captured its itinerary in a series of chronological snapshots: in the first, an ordinary Kenyan in a neat suit and shined shoes stands sulking under white colonial rule. In the second, a free man under Kenyatta leaps for joy, but his suit is beginning to look distinctly tatty. By the Moi era, the emaciated *mwananchi* is crawling, not walking. His suit is in tatters, he has lost his shoes, and, eyes crazed, he is begging for alms. The statistics made the same point, in drier fashion: living standards in the independent, sovereign state of Kenya were actually lower than when the hated British ruled the roost.

Kenya might well boast, by African standards, a large middle class, but the gap between that group and those eking out a living in its teeming slums was the stuff of revolution. 'Kenya is now one of the most disappointing performers in sub-Saharan Africa,' ran an editorial in my own newspaper, the *Financial Times*, the day after the 2002 election. 'There is barely an economic or social indicator that does not testify to the country's decline.'[2] Given that Kenya had never experienced a civil war, never been invaded, and had started out with so much in its favour, the fault must lie elsewhere. And everyone agreed where: in a system of corruption and patronage so ingrained, so greedy it was gradually throttling the life from the country.

Whether expressed in the petty bribes the average Kenyan had to pay each week to fat-bellied policemen and local councillors, the jobs for the boys doled out by civil servants and politicians on strictly tribal lines, or the massive scams perpetrated by the country's ruling elite, sleaze had become endemic. 'Eating', as Kenyans dubbed the gorging on state resources by the well-connected, had crippled the nation. In the corruption indices drawn up by the anti-graft organisation Transparency International, Kenya routinely trailed near the bottom in the 1990s, viewed as only slightly less sleazy than Nigeria or Pakistan. From the increasingly strained relations between the country's tribes to the rising anger of its prospectless youth, Kenya exemplified many of Africa's most intractable problems.

Which is why so many eyes now rested on Kibaki and his NARC government. If they could get it right on corruption, if Kenya could only find its way, then perhaps there was hope for the rest of the continent. Post-apartheid South Africa, post-military Nigeria and a revived Kenya could come to form the three points of a geographical triangle of success establishing Africa on firm, unshakeable foundations.

The first announcements the new president made after unveiling his cabinet continued to send out the right signals. A brand-new post

– Permanent Secretary in Charge of Governance and Ethics – was being created, Kibaki said. This anti-corruption champion, Kenya's version of Eliot Ness, would run a unit working out of State House and enjoying direct access to the president's office. And that key job was going to someone who seemed tailor-made for the role, a young, energetic Kenyan who had dedicated his considerable talents to the fight against graft. He just happened to be an old friend of mine.

I had known John Githongo since moving to Nairobi in the mid-1990s, when he was an up-and-coming columnist and I was the *Financial Times*'s Africa correspondent. Kenya's newspapers were good, among Africa's best, but their columnists suffered from parochialism. They didn't travel the region, they had little sense of Africa's position in the world, they sounded uncertain when tackling international issues. John, who wrote a think piece for the *EastAfrican*, a business weekly owned by the Aga Khan's Nation Media Group, was different. He had studied abroad, had travelled his own continent and had a sound grasp of geopolitics. His vision was sophisticated, his instincts compassionate, and he had the good journalist's ability, using colourful anecdote, to make complex arguments accessible to the ordinary reader. Limpid and articulate, his columns commanded one's attention, like a clear voice carrying across a room of cocktail chatterers.

I asked him to lunch. A giant walked in. 'The Wrestler', he would later be dubbed by staff at Kroll, the London-based risk consultancy group. But for most of us, the tag that automatically sprang to mind on first meeting was 'the Big Man'. Standing well over six feet, he had girth as well as height, the fifty-eight-inch chest and massive shoulders of the gym habitué, the V-shaped silhouette of a comic-book superhero. He was a gift to any caricaturist, but this exaggerated outline was built of muscle, not fat – squeezing one of those rounded shoulders in greeting was like kneading a well-pumped football: the fingertips left no impression behind. It was a bully's physique, but no bully ever walked with his tentative, splay-footed step, the step of a man anxious not to tread on smaller mortals milling below. He wore his hair very short, snipped to virtual baldness to reveal a bull neck and a formidable jaw, something of a family feature. Faces in the Githongo family, I would later discover, had the all-weather implacability of Easter Island sculptures. He was a Kikuyu, but his enemies would later claim that he didn't look as though he belonged: too big, too tall, too dark. He photographed supremely badly – I never saw a photograph of John that made him look anything but stolid, loutish, slightly thick. Still only in his thirties, he looked older than his years, thanks to the receding hairline, deep baritone and seeming gravitas. In fact, John was prone to fits of the giggles. An inveterate conspiracy theorist, intrigued by tales of plots and subterfuge, he loved a meaty gossip. 'Is that SO? Is that SOOO?' he would whisper in fascination on being passed a nugget of clandestine information, mouth forming a round 'O' of wonder, eyes growing big as he dwelt on some VIP's quirk of character, or the little-known story behind some public political clash. And it was hard to think of anything, or anyone, that *didn't* interest his questing mind. Blessed with insatiable curiosity, he gobbled up experiences and insights in the same way he embraced new acquaintances.

He returned my lunch invitation by asking me along to a meeting of the Wednesday debating group, a serious affair where earnest young men in suits discussed topical issues. After that we saw each other only sporadically, but it always felt like time well spent. There was no hint of anything romantic, nor would there ever be. John was simply one of the most intellectually impressive young Africans I'd met, and each encounter left me optimistic for the country's future.

These were the days when Kenya's opposition parties were trying to get a constitution weighted in Moi's favour changed. Mounted on horseback, the dreaded GSU, Darth Vader-like in shields and helmets, charged supporters who dared defy a ban on public rallies, lashing out with truncheons and pick-axe handles. I remember venturing out behind John on a day when the oniony smell of tear-gas was still wafting along the city centre's deserted streets, and noting how nimbly he darted along the pavements and peeked round corners. His caution made me nervous. If a man his size was worried about running into the GSU, I thought to myself, then I should really watch out.

I began quoting John in my articles. Other Western journalists were also discovering him. Soon the name 'John Githongo' was cropping up in more and more media stories as a pundit. Then he'd left journalism to revive the local branch of Transparency International, an organisation established by his own father and a group of like-minded Kenyan businessmen disillusioned with Moi. He had found the perfect platform from which to hold a morally bankrupt government to account. John knew instinctively how the media could be mustered and energised to contribute to the democratic reform process – he'd been a journalist, after all. TI's carefully researched reports finally quantified Kenya's amorphous corruption problem, giving the media something solid to get their teeth into. John cultivated contacts, put out feelers, and sailed so close to the wind he found himself being offered political asylum by concerned fellow delegates after telling an anti-corruption conference in Peru what he knew about president Moi's portfolio of investments. The foreign governments who funded many of TI-Kenya's activities loved it. Here was concrete proof of how donor aid, cleverly directed, could bolster accountability in Africa.

While working at TI, John was also in discreet contact with the Kibaki team. He'd kept that side of things quiet, for the organisation was officially neutral, and had to be seen to remain above the political fray. But when Kibaki's aides approached, asking for concrete suggestions on how to build the opposition's anti-corruption strategy, he could hardly refuse. And in truth, at this stage in Kenya's history it was almost impossible to imagine that any idealistic young Kenyan could fail to wish NARC anything but success in the forthcoming contest.

In the wake of the 2002 inauguration I tracked John down with a fellow journalist, keen to hear his thoughts. We found him in frenetic mode, simultaneously hyped, exhilarated and exhausted. He had been part of the election-monitoring effort pulled together by the human rights bodies and advocacy groups that constituted Kenyan civil society, and was fielding a series of calls from reporters in search of quotes, repeating the same phrases again and again. Halfway through the conversation, he revealed another reason why he was so distracted. The Kenyan businessmen who sat on TI-Kenya's board, old friends of both his father and Kibaki, had been in touch. 'The *wazee* [old men] have put my name forward as someone to lead the fight against corruption.' His laugh was half-embarrassed, half-excited. 'It looks as though the new team is going to offer me a post in government.'

My heart sank. I could see exactly why any new government would want John. No Kenyan could rival his reputation for muscular integrity, or enjoyed as much respect amongst the foreign donors everyone hoped would soon resume lending. In co-opting him, the incoming administration would be neatly appropriating a high-profile symbol of credibility, proof personified that it deserved the trust of both the *wananchi* and its Western partners. But I remembered all the other shining African talents I'd seen warily join the establishment they had once attacked, persuaded that finally the time was ripe for change, only to emerge discredited, beaten by the system they had set out to cure.

'Don't take it,' I said. 'You'll lose your neutrality forever. Once you've crossed the line and become a player, you'll never be able to go back.'

He listened, but my advice, it was clear, was being given too late. Effectively, he explained, he wasn't being given a choice. The old guys – Joe Wanjui, former head of Unilever in Kenya; George Muhoho, founder member of Kibaki's opposition Democratic party; and Harris Mule, former permanent secretary at the finance ministry – had done the deal in his absence, taking his acquiescence as read. He'd gone round to Wanjui's house and found the *wazee* drinking champagne, celebrating the forthcoming appointment. They had ribbed the young man over the fact that he probably didn't even own a suit for his meeting with Kibaki, offering to lend him one. 'They'd all cooked it up together. I drove away stunned. It was a great honour.' In later years, he would think back over that day and detect an unappetisingly sacrificial element to the whole episode. These men he had grown up with, who had known him when he was nothing but a small boy running around in shorts, had trussed him up and delivered him to his fate.

But it was obvious that John was more than a pawn in a deal done by his father's friends. He was the kind of man who believed it was up to every Kenyan – especially to someone blessed with his education and social advantages – to pull the country out of the mire. He had dedicated his brief career to fighting corruption. Now along came an administration that had won an election promising to do just that. It was asking for his expertise, inviting him into the inner sanctum, and he knew in his heart that there probably wasn't a single Kenyan better placed to wage that campaign. How could it be legitimate to criticise if, when you were explicitly asked to quit the sidelines and join the fray, you refused? 'We discussed whether he should take it and concluded he didn't have a choice, morally speaking,' remembers economist David Ndii, who had worked alongside John at TI. 'If he didn't, he would always wonder if he could have made a difference.' There comes a time in a man's life when fate offers him a chance to do something significant. It is rarely extended twice. Accepting the job was not just an exciting career opportunity, it was a patriotic duty.

Leaving John that day, I felt a deep tinge of melancholy. Working in Africa, I'd grown accustomed to compromised friendships, relationships premised on wilful ignorance on my part and an absence of full disclosure on my friends'. When visiting a former Congolese prime minister, sitting in a villa whose bougainvillea-fringed gardens stretched across acres of prime real estate, I knew better than to ask if his government salary had paid for all this lush beauty. Staying with a friend in Nigeria, whose garage alone dwarfed the family homes of many Londoners, I took it for granted that his business dealings wouldn't stand up to a taxman's scrutiny. And when I shared a beer with a Great Lakes intelligence chief befriended in a presidential waiting room, I knew that one day I'd probably come across his name in a human rights report, fingered as the man behind some ruthless political assassination. Life was complicated. The moral choices needed to rise to the top were bleaker and more unforgiving in Africa than those faced by Westerners. It was easy for me, born in a society which coddled the unlucky and compensated its failures, to wax self-righteous. I had never been asked to choose between the lesser of two evils, never had relatives beg me to compromise my principles for their sakes, never woken to the bitter realisation that I was the only person stupid enough to play by the rules. If I was to continue to like these men and women – and I *did* like these men and women – it was sometimes necessary to focus on the foreground and wilfully ignore the bigger picture.

But not with John, never with John. Through the years of knowing him, I had never caught a glimpse of any sinister hinterland, territory best left unexplored, and God knows I had asked around. What you saw seemed to be strictly what you got, and he was the only one of my African friends of whom that felt true. I looked at him that day and thought: 'Well, that's over. In the years to come, I will pick up a Kenyan newspaper and spot an item in a gossip column about his partnership with a shady Asian businessman, the large house he is having built in a plush Nairobi suburb. Then there'll be a full-length article, a court case in which the judge finds against him but which goes to appeal, so I'll never know the truth. And one day, I'll be chatting to someone at a diplomatic party who will say: "John Githongo – isn't he completely rotten?" and I'll find myself nodding in agreement . . .' Oh, I would still like him – who could not? But what had once been clear-cut and simple would have become qualified and murky. And already I mourned our mutual loss of innocence.

There was one last hoop to jump through before his appointment was confirmed – an interview with the man who had just become Kenya's third president. At that first encounter on 7 January 2003, watched over benevolently by the *wazee*, his three mentors, John listened, humbled, overawed, as Kibaki outlined his ambitions and expectations. But he plucked up just enough courage to make a remark that went to the heart of the matter. If his time at TI had taught him one thing, he said, it was that since corruption started at the top, it could only effectively be fought from the top. 'Sir,' he told the president, 'we can set up all the anti-corruption authorities we want, spend all the money we want, pass all the laws on anti-corruption, but it all depends on you. If people believe the president is "eating", the battle is lost. If you are steady on this thing, if the leadership is there, we will succeed.'

Among the many calls John received in those hectic days, as excited friends rang to congratulate him, one was more sobering than the rest. It was from Richard Leakey, the palaeontologist who, after years in opposition, was taken on by Moi in the late 1990s to reform Kenya's civil service. Leakey was no stranger to adversity – he had been hounded by the security forces, bore the scars on his back from a vicious police whipping, had lost his legs in a plane crash some suspected of being a botched act of sabotage. An experienced scrapper, his efforts to clean up the public sector had nevertheless eventually been rendered futile by Moi's Machiavellian strategies. 'If you can pull it off, wonderful,' Leakey told John. 'But be careful. This is a tough one.' The appointment was announced in the following days, to much media fanfare.

After that, we rarely met. I was busy writing a book in London, John was a man in a hurry. After a lull, I started getting the occasional, worrying bulletin: he had made some powerful enemies, and travelled around Nairobi with two bodyguards; new scandals were surfacing; John had been moved sideways, then reinstated. That didn't sound good. It got worse: a journalist friend returning from Nairobi said John had told him that 'if anything happened' he had left instructions for both of us to be sent certain packages, an ominous sign if ever there was one. And his hitherto unblemished reputation was taking its first hits. Nairobi's chattering classes were complaining that the anti-corruption chief wasn't delivering. Whether through ignorance or impotence, they said, he was complicit in the new government's misdemeanours. He was going down the route the cynics had always traced for him, from superhero to flawed mortal.

Then, on a visit to Kenya in late 2004, John joined a meal I was having in a French restaurant with four Western correspondents, veteran Africa writers all. His arrival was a welcome surprise, for John

– always prone to the last-minute cancellation – had become outrageously unreliable since joining government, as notorious for his no-shows as a Hollywood diva.

'So, John, when are you going to resign?' asked one of my colleagues, and John chuckled ruefully, shaking his head in defeat.

As we prepared to leave, I turned to him on sudden impulse. He had not said as much, but under the ebullient cheerfulness that was his customary public face, I thought I glimpsed a certain dismay. He seemed buffeted, a man no longer in control of his destiny.

'I've just moved into a larger flat in London, John, with a separate guest room. If you ever need a base' – the phrase 'bolt hole' was on the tip of my tongue – 'somewhere to rest up, just give me a call.'

The response came a few months later. A call from Davos, where John was attending the World Economic Forum. 'I was wondering if I could take you up on that offer of a room?' He gave no hint of how long he planned to stay or why he needed a place for the night when presumably, as a government VIP, he enjoyed the pick of London hotels. When he called again, this time from Oslo, where he was attending a conference, I asked whether his visit was something I could mention to journalist friends in London, always keen to see him. 'Er . . . Probably best not. If you don't mind, just keep it to yourself for now.'

Something, clearly, was up. And on the morning of 6 February 2005, when the capital was wrapped in a cold white cocoon, he arrived on the doorstep of my London flat, let in by a genteel elderly lady from down the hall who seemed, to John's quiet amusement, to find nothing remotely suspicious about a huge black man in a KGB-style black leather jacket, herding a pile of luggage so large it was clear that this would be no weekend stay. As he deposited the various bags in my guest room, which suddenly looked very small and cramped, John's mobile phones trilled and vibrated, like a chorus of caged starlings. How many did he actually have: three? four? more? He asked for a glass of fruit juice, took a deep breath, and gathered his thoughts.

'One of the first things I need to do,' he said, 'is resign.'

He was on the run, he told me. In best espionage style, he had summoned two taxis to the London hotel where he had been staying with Justice Aaron Ringera, head of Kenya's Anti-Corruption Commission, paid one to drive off in any direction and taken the second. Whatever I might have fondly liked to think, his appearance on my doorstep at this moment of crisis was scarcely a tribute to the intimacy of our friendship. Quite the opposite, in fact. He was there precisely because so few people in Kenya knew we had ever been friends.

'They told me it was *them*,' he said, pacing the floor. 'These ministers, my closest colleagues, sat there and told me to my face that they, *they* were the ones doing the stealing. Once they said that, I knew I had to go.'

2

An Unexpected Guest

'If you're walking in the savannah and a lion attacks, climb a thorn tree and wait there for a while.'

Kamba proverb

He came bearing toxic material. A nervous tremor scurried along my spine as he explained that he had done the unthinkable, wiring himself for sound in classic police informer style, taping the self-incriminating conversations of the ministers who were supposed to be his trusted workmates. The explosive contents of those recordings had been systematically downloaded onto his computer, which now sat quietly in my spare bedroom. 'It might be an idea,' he said, 'for me to find a third party to take the computer while I work out what I'm going to do.'

Suddenly, I was plunged into an unfamiliar world – of covering my tracks, watching what I said. In this world of subterfuge, even the simplest procedure grew vastly complex. Sitting at my computer, John wasted no time in typing out his resignation letter. He drafted it slowly and carefully. While he did not want to give anything away that might constrain his actions later on, he was also determined to make it clear to the careful reader – and he knew State House, the intelligence services and the media would be analysing every word – that he was not leaving happy in the knowledge of a job well done. There would be no 'spending more time with my family' clichés. The circumstances of his resignation alone, announced on a one-way trip into exile, must at this stage do the rest for him, sending a damning message about the true nature of the NARC regime.

He was in a hurry to cross that Rubicon; the letter needed to be faxed immediately to State House. But whose fax machine to use? If I used my own, his location would immediately be revealed. My parents' fax would be no better – given my family's unusual surname, it would immediately lead anyone with half a brain back to me. Nearby Camden Town was full of little newsagents willing to fax documents for customers. But in my experience, most were run by sulky Asian shopkeepers who had no truck with international calls. In any case, a Camden Town telephone number would once again point Kenyan investigators in my general direction. In the end, despairing of getting it right, I walked into an independent bookshop I regularly patronised and asked the owner – a laid-back, gently humorous man who had done me many favours over the years in return for my loyal custom – to fax the letter, hoping he wouldn't notice its recipient ('President's Office, State House, Nairobi') as it passed through his hands. He was a Jewish émigré's son. His father had fled the Nazis and saved himself from the concentration camps; I told myself he should understand about life lived under the radar.

The resignation was splashed across the front pages of Kenya's newspapers in three-inch capitals the next day, the only topic of conversation on the FM radio stations, morning chat shows and Kenyan websites. 'STATE HOUSE SHOULD BE CONDEMNED NOW! PARLIAMENT SHOULD BE CLOSED! TAXPAYERS, STOP PAYING YOUR TAXES, IMMEDIATELY!' ran one typical blogger's entry. The one that followed quietly summarised the national feeling: 'Shit.' Even the international media ran hard with the story, realising this was an event likely to damage relations between the Kenyan government and its new-found foreign friends. After only two years in his post, the living, breathing symbol of Kibaki's good intentions had thrown in the towel, the shining white knight had fallen off his horse. Who could remember a similar event in African, let alone Kenyan history? Permanent secretaries *never* surrendered their jobs, they were either ignominiously sacked or, if they were lucky, allowed to present token resignations. Had John resigned in Nairobi, it would already have been remarkable. The fact that he had chosen to do so from self-imposed exile – indicating he believed his life would be in danger if he stayed in Kenya – made it one of the hottest African stories of the year. In Britain's House of Commons, a Labour MP tabled a private member's motion expressing his 'deep concern', while the missions of the United States, Britain, Canada, Germany, Sweden, Norway, Japan and Switzerland called in a joint statement for Kibaki to take swift action to restore his government's credibility.

In the days that followed, the Kenyan government mounted a quiet manhunt. As Special Branch descended on John's house in Nairobi – 'It was just like the old days,' a friend who lives in the same district later told me, 'with police cars drawing up in the night, neighbours woken, dogs barking' – staff at the Kenyan High Commission in Portland Place scoured London. They checked the addresses of John's known friends, people he had grown up and gone to school with. Nothing. They canvassed the roads around Victoria Station, an area of cheap lodgings patronised by Africans who can't afford the top hotels. No luck. No one thought to check the home of Michela Wrong, former Africa correspondent of the *Financial Times*.

But the pitfalls inherent in a life of deceit were swiftly becoming obvious to me. I'd had colleagues who had crossed the invisible line of journalistic neutrality and become part of their own story, giving succour to African asylum-seekers, paying their legal fees, sneaking money and papers across borders. But it had never happened to me. And, it turned out, I wasn't much good at this stuff. David Cornwell, better known as John le Carré, a veteran of subterfuge, later told me that a cover story only works if prepared ahead of time, its structure and corroborative detail laid down well in advance. But I had had no time to prepare my 'legend'. I was reacting on the hoof, and within hours, not days, I was tripping myself up.

The main problem was that I didn't want to mention John's presence over the telephone. I had always vaguely assumed that, whatever the British authorities might say in public, any intelligence service worthy of its name routinely bugs its small resident community of journalists, particularly in the wake of 9/11. British intelligence, I knew, had an information-sharing arrangement with its Kenyan counterpart. If asked by Kenyan intelligence to help track down a missing anti-corruption chief in London, would the Brits refuse? I wasn't counting on it. So how to explain to the various girlfriends of mine who rang, expecting an intimate natter, that this wasn't a good time for our usual gossip, without explaining why? They immediately sensed the awkwardness in my voice, and the less I divulged, the more curious they became. 'Do you have someone there? Why are you being so secretive? What's going on?' If I'd been listening in on those conversations, my limp 'I'll explain later' would immediately have alerted me to the fact that something had changed in the Wrong household.

I found myself in a similar predicament when John asked if I could book lodgings for a Kenyan contact passing through London who he needed to meet. Suspicious of everyone in these tense, early days, he preferred his visitor not to know where he was staying, which raised the awkward question of how to pay for the room. If the guest were to ask at reception who was covering his bill, my credit card details would give the game away. If I went to the bed and breakfast in person and paid cash, I risked making myself memorable by that very act. I rang my brother-in-law – same family, different name – and asked if he would mind charging a room in north London to his card. 'Er, I could, but why can't you just pay for it yourself?' 'I can't explain why now, but there's a good reason,' I muttered. 'Well, if you're not going to tell me, I don't want to be any part of this,' he said, turning unexpectedly priggish. The exchange made me scratch tentative plans to hand John's 'hot' computer to my in-laws for safekeeping. It was surprising how little you could get done, once frankness was ruled out.

And then there was the outright lying. The Kenyan government wasn't the only organisation trying to track John down. Even Kenyan bloggers momentarily turned amateur sleuths, swapping notes on their websites as to which London hotels had confirmed he wasn't a guest. There were calls from the BBC World Service, emails from Kenyan journalists who had caught a whisper of something in the air; an ambassador left cryptic messages on my answerphone, sending his best wishes to 'our mutual friend'. Did I happen to know, the journalists asked with deceptive casualness, where they might get hold of John Githongo? It really was most urgent that they talk to him. He might be in possession of some very interesting information. As John pottered around in the background, doing his laundry and preparing his lunch – no macho African nonsense about him – I'd breezily debate his possible whereabouts and motivations with hacks I'd known for decades, hoping he wouldn't blow his cover by saying anything in that distinctive baritone.

In theory, I should have been pestering him for an interview myself. In fact, I held back. While I was clearly sitting on a fabulous story – Africa's Watergate, by the sound of it – sitting John down with a notebook and tape recorder would have felt like a cheap trick, his host joining the manhunt rather than offering the safe haven he clearly desperately needed. Perhaps a less noble instinct also lay behind my uncharacteristic discretion. In the world John had entered, it seemed, knowledge made you a marked man. Once I too knew whatever it was he'd learnt, maybe I would face the same predicament. I wasn't sure I was ready to catch that particular infection. So I mentally stored the nuggets of information that came my way, while allowing the overall picture to escape me. He talked of ministers, he mentioned a naval vessel, the words 'Anglo Leasing' came up repeatedly. But he never joined up the dots. I wondered, once or twice, what I would actually be able to say to the police if something sinister happened to him. I'd have no coherent tale to tell, and they would surely refuse to believe that an intelligent journalist, harbouring a political fugitive, had never bothered to fit the various pieces together.

Out on the street, I scanned black faces with a paranoid new attentiveness, trying to spot the undercover Kenyan agent attempting to blend in. But Camden has an awful lot of Africans living in it. From my new and wary perspective, almost *everyone* looked suspicious. At night I lay in bed, pondering how far the Kenyans might go. I was aware that I was thinking exactly like a character in a thousand Hollywood thrillers, but this fear was surely rooted in cold logic. I ticked off the various factors on my personal risk assessment. Did the material on John's computer have the potential to bring down a government? From the little he'd sketched out, yes. Were the reputations and livelihoods of Kenya's most powerful men – possibly the president himself – at stake? It seemed so. Did Kenya have a history of ruthless political assassination? Absolutely – I could reel off the names: Pio Pinto, Tom Mboya, J.M. Kariuki, Robert Ouko, Father Kaiser – and those were only the most notorious cases. Kenya had always been a venue for the well-timed car crash, the fatal robbery in which both gangster and high-profile victim conveniently lose their lives, the inquiry that drags on for decades and then sputters out without shedding any light on what had really happened.

Were the stakes this time high enough to be worth killing a man? Clearly, John believed so, otherwise he wouldn't have fled. So the only question that remained, from a selfish point of view, was whether the Kenyans would be foolhardy or desperate enough to try something on British soil. Which meant my flat. After triple-bolting my front door – I was glad now that I'd bought the most expensive lock on the market when I moved in – and slotting the chain into position, I'd fall asleep in the early hours, stressed and fraught. In my dreams, a huddle of burly figures in formless grey overcoats with blurred, dark, hatchet faces, battered their way in to shoot us both in our separate rooms.

In the morning, after a restless night, I'd wake feeling embarrassed by my melodramatic thought processes. If I was finding John's stay a bit of a psychological ordeal after only a few days, what must it be like for him? How had he endured the last few years, living with that anxiety day by day? Yet he seemed astonishingly cool. For the most part he ignored his collection of mobile phones as they constantly vibrated and shrilled. Occasionally he'd pick one up, disappearing into his room to hold a quiet, intense conversation in Gikuyu or Kiswahili. But usually he would just look at the display, check who was trying to make contact, then put the handset down. The one that rang with most persistence was his line to State House.

'It's very interesting,' he mused. 'They haven't cut off my State House mobile phone. My safe in the office hasn't even been opened. And my secretary is still at her post.'

'It's their way of telling you that you can still go back,' I suggested. 'They're saying, "It's not too late, the lines are still open."'

Yet even by that stage, I had begun to recognise what constituted signs of stress in the Big Man. His booming, seemingly carefree laugh was the equivalent of most people's titter – a sign of tension, not relaxation. The more nervous he became, the more heartily he laughed. He wasn't sleeping well either – I gave him some of my sleeping pills when he mentioned the problem – and his mental fatigue was evident in his tendency to tell me the same things over and over again. His sentences were like ripples on the surface of a pool – they gave a hint of the thoughts churning obsessively in the depths below. I could guess what those might be: How on earth had it ever come to this? Was this the right path? Where did he go from here?

The best way of relieving the stress was exercise. John was the kind of dedicated workout enthusiast who knew which machine targeted exactly which muscle group. One of the first sorties we made from my flat was to tour the local area scouting out which gym had the best weight-training facilities. Working out – a three-hour process – was not just a hobby, he *needed* it, needed to feel the adrenalin coursing round his body if he was to stay focused and sane. Other men might have started working their way through my drinks cabinet, but my fridge filled up with cartons of fruit juice. John, iron-disciplined in this as in so many things, had turned teetotal during his time in State House, when he had noticed that winding down from a stressful week with a bottle of whisky had become a habit, and that the habit was becoming increasingly hard to break. It was typical of him that he wouldn't let himself slip back, not even now, when he had the best of excuses for needing the odd stiff drink.

His other recourse was religion. Having spent so much time in Britain, John had registered the scepticism, if not downright antagonism, of his European acquaintances when it came to matters religious. His Catholic faith was something he never talked about with his *mzungu* friends, I noticed, turning instead that side of himself with which they felt most at ease. Only the Virgin Mary medallion around his neck and the rosary ring on his finger – one metal bobble for each Hail Mary to be recited, removed only during weightlifting – gave the game away. But one of his last visits before leaving Nairobi had been to call in on the Consolata Shrine, where troubled minds went in search of solace. And in those fraught early weeks in London he did a lot of praying.

As he quietly came and went, reuniting with girlfriend Mary Muthumbi – an advertising executive who flew to London to see him – officially registering his presence with a Foreign Office that expressed only polite interest, a silent question mark was forming. Fleeing the country, in a way, had been the easy part. What, precisely, was he going to do next?

As far as I could see, there were only two options. Option One: Leave government employment and keep quiet. Give the tapes and computer material – your insurance policy against assassination – to a British lawyer, along with firm instructions that should anything happen to you, they will be released to the press. Make these arrangements clear to those in power, and assure them you will never give another media interview in your life and will never go into politics. Work abroad, go into academia, get married to your long-suffering girlfriend and wait for the affair to die down. Eventually, maybe five years down the line, you will be able to return to Kenya, and while ordinary folk will look at you with a certain cynicism and think, 'I wonder what he knew?', most will respect your discretion and commonsense. No man can single-handedly transform a system, and you will be joining the ranks of former civil servants with clanking skeletons in their cupboards. Your conscience may occasionally trouble you, and you will have to acknowledge that you tried and failed. But you will have got your life back.

Option Two was bleaker, more dramatic, and fitted straight into that Hollywood thriller genre. Lance the boil, go public. Blow the government you once passionately believed in out of the water and say what you know. People who matter may hate you for all eternity. You may never be able to go home again, your family and friends may suffer by association, your colleagues may regard you as a traitor, but you will have done the right, the upstanding thing, and lived up to the principles that have governed your life. You will have shown the world that others may do as they please, but as far as you are concerned, 'Africa' and 'corruption' are not synonymous.

Most journalists, I suspected, would urge John to choose Option Two – it made for a fantastic story. I urged him to choose Option One. Those journalists would not have to live with the consequences. My old friend, it seemed to me, had already done his share, and his country's fate was not his burden to shoulder alone.

Initially, he'd planned a press conference. The speculation and allegations being published in the Kenyan press irked him, he said, and he felt he owed the Kenyan public an explanation. I quailed at the thought of the bun-fight that would follow.

'If you're going to hold a press conference, you have to be absolutely clear in your mind what you're prepared to say. Are you going to spill the beans now? Are you ready to explain what actually happened?'

'No, not yet.'

'Then don't do it. The most infuriating thing you can do to journalists is to hold a press conference and say nothing. It'll drive them crazy. They'll either force you into making admissions you don't intend or rip you to shreds for wasting their time.'

Another idea he considered, urged on him by the few friends in London who were gradually discovering his whereabouts, was to record an 'in the event of my death' videotape in which he named names and explained his departure. If he were killed, it would remain as devastating testimony. He toyed with the idea, but held off once again. Perhaps he was wary of creating such an incendiary tape – who could be trusted to keep such red-hot footage under wraps? But it was also a question of strategy. John's modus operandi, perfected over the years, was to painstakingly think through every eventuality, harvesting the insights of well connected insiders, visualising every possible scenario before moving to action. 'I try and dot all the "i"s and cross all the "t"s. I do this excessively, it's been my style throughout. And then, when I move – BOOM!' The approach slowed him down, but he needed to feel he had set his intellectual house in order. If he taped an interview so early on, he'd be skipping the methodical preparation of the ground that felt like a necessity.

A fortnight later, with the key questions unanswered, John moved out. He headed first to the home of Michael Holman, another British journalist whose friendship with him was as little known as my own, and then to a scruffy flat next to a north London fish-and-chip shop.

I didn't like to admit it, but his departure came as both anticlimax and relief. There was no denying that my brush with a man at the vortex of a major political crisis had provided me with a vicarious thrill. But there had been a few close shaves, close enough to make me uncomfortable. My parents' flat happened to be situated around the corner from the Kenyan High Commission. Once I'd caught a bus that stopped just outside the building and two Kenyan women employees, leaving for the day, had boarded after me. To my alarm, they had ridden all the way to my bus stop. These women, who would certainly know John in his official capacity and recognise him if they bumped into him on the street, lived in my local area. Another time, John had been using a local cyber café and a Kenyan customer had suddenly started chatting to him in Kiswahili. It was not clear whether he'd been recognised, or this was just a case of one East African being friendly to a fellow abroad. Had I been working for Kenyan intelligence, I would have simply toured all the London gyms with good weightlifting facilities, asking if a large black Kenyan had recently signed up. But John warned me that his emailing would be the activity that would eventually lead his pursuers to him. By analysing the emails he sent back to Nairobi, he said, it would be possible to eventually work out the geographical location of the terminals he'd used. Sure enough, a Kenyan newspaper editor who liked to show off the extent of his intelligence links would later drawl when I walked into his Nairobi office, 'So I hear John Githongo has been holed up with Michela Wrong and Michael Holman in London.'

A few weeks after finding his own place, waiting on a London Underground platform, John realised he was being followed by two middle-aged Kenyan men who looked exactly what they almost certainly were: undercover agents. He sprinted down a passageway and hopped onto a train to lose them. Then one day, emerging at his local tube station, he was confronted by a Kenyan man, standing coolly watching him, making sure John registered his presence. They had tracked their prey down to his lair, and were showing off the fact that they knew where to find him.

Yet they did nothing. There was no attempted break-in to verify what, if any, material he held in his new lodgings, no raid to confiscate the incriminating laptop – still in his possession and containing plenty of unbacked-up material – no overture, no whispered threat, no attempt to lure him back to Kenya. They were hanging back, waiting. Waiting for what, exactly? Presumably for the same thing as the rest of us: waiting for the Big Man to make up his mind.

He moved yet again, this time to Oxford's St Antony's, a college with a history of offering sanctuary to those in political hot water. Professor Paul Collier, an expert on African economies, had come to the rescue with a not particularly demanding senior associate's post on its East African Studies programme. It was exactly the kind of academic berth John needed at this juncture, offering him accommodation, a work space and – crucially – the time in which to gather his thoughts.

One of his first acts there was highly symbolic. Just as his government experience had been at its sourest, he had been named Chief of the Burning Spear, Kenya's equivalent of the Order of the British Empire. Coming when it did, the award had felt part consolation prize, part bribe. Now he arranged for it to be sent to an old Kenyan friend, Harris Mule. Mule, a former permanent secretary at the finance ministry, had been a loyal civil servant who had refused to play the political game. When he had fallen into disfavour, he had quietly accepted his fate. John had consulted him when things got difficult, drinking in his wise advice. Now he sent Mule a medal he believed he himself did not deserve, and which Mule should have been awarded decades ago. If State House was ever made aware of that small gesture, it would have been well advised to take notice. There was a touch of the boat-burning about it.

Ensconced in his new lodgings, John was nothing if not methodical. Now that he had caught his breath, it was time to pull everything together: the contents of the diaries he had kept throughout two years in office – well-thumbed, numbered black notebooks transcribed in neat fountain pen, the sloping handwriting squeezed as close as possible to make maximum use of space – the documents he had copied and quietly sent abroad, the digitalised tape recordings downloaded onto his computer. If he was ever to make head or tail of it, all this information needed to be scanned, logged, written up and placed in some logical order. To date, he had turned down every interview request, made no statements, held no press conference. He had marked his fortieth birthday, that psychologically significant moment in a man's life, with the start of a new, uncertain existence in a foreign country. All paths still lay open to him. But he would only know what to do next once he had understood exactly what had happened to him. And to Kenya.

3

Starting Afresh

'Youth gives all it can: it gives itself without reserve.'

JOSEMARÍA ESCRIVÁ, founder of Opus Dei

There's a certain sameness about presidential lodgings in Britain's former African colonies, and Nairobi's State House, the former colonial governor's residence, is no exception. Fall asleep in the waiting room and on waking you could, in that bleary moment of confusion, think yourself in State House, Zambia; State House, Tanzania; or State House, Uganda. Behind the white-pillared porticoes they present to the world, these buildings are resolutely dowdy, content to remain stubbornly out of touch with modern trends in interior design. No stark minimalism here, no streamlined vistas, no clever games with reflection and light. The décor is dark wood panelling, chintz sofas, red carpets and thick velvet drapes. The taste in pictures will usually be execrable: an anaemic watercolour of an English country scene, an uplifting motto urging the reader on to greater Christian efforts, an oil portrait of the incumbent so approximate it could have been sat for by someone else entirely. The carpet will be worn through in places, a clumsily carved piece of animal Africana will take up a great deal of space. The overall impression is of a dusty members' club crossed with a gloomy British country pub, and the effect is to make those indoors pine for the fresh green of the formal gardens outside, the only real area of beauty.

One of the peculiarities of Nairobi's State House is that it is invisible from the road, the only hint of its existence a formidable checkpoint and a challengingly high metal fence. Puzzlingly, this fence has repeatedly failed to do its job. In the wake of Moi's unceremonious exit, several solo intruders were discovered wandering the presidential grounds in the early hours. One was an Australian tourist, another a Ugandan. Arrested by the GSU, they could not satisfactorily explain what they were doing on the premises. After some initial headlines, they were never heard of again. Word spread amongst Nairobi's more superstitious residents. These mysterious visitors had been able to pass through State House's supposedly impregnable fence, then evaporated into thin air, because they were not men at all, but spirits. Jomo Kenyatta had refused to spend a single night in State House, convinced it was haunted by vindictive ghosts of the white administration. Moi, it was now said, had also left a malign parting gift behind, an evil genie, a curse which explained not only these night-time visitations but the variety of misfortunes – from Kibaki's near-fatal car crash to the death of his first vice president – that befell its new incumbents.

It was here, in an old bedroom converted into a study, that John set up base in early 2003. On his desk he placed a framed picture, a present from Bob Munro, a Canadian friend who ran a slum-based soccer-club scheme. It was a copy of a Charles Addams cartoon, showing a skier whose parallel tracks in the snow surreally divide and rejoin on either side of a pine tree. 'That's going to be you,' Munro joked, anticipating the impossible demands that would be made on the future civil servant. John had initially established an office outside the main building, within the State House compound. The president was having none of it. 'No, no, no, I want you inside this building,' Kibaki had said, insisting that the newly appointed anti-corruption czar should be virtually within shouting distance of his own office – just two doors and a foyer separated them. 'Don't brief anyone but me, don't bother making appointments, just check that I'm free and come straight in.' It seemed the president had taken John's message on board.

That physical proximity alone ensured John extraordinary influence. In a strong presidential system, being in a position to brush against the head of state in a corridor is worth a score of weighty-sounding titles. Whatever John's nominal grade, being granted free licence to update the president whenever he wished effectively placed him above many cabinet ministers in the pecking order. And Kibaki was true to his word. By the time he left, John calculated that he had given his boss sixty-six briefings, some of them stretching over two or three separate meetings. That walk-in access made him a player of huge interest to anyone wanting to cut through the layers of bureaucracy to reach the core of power. 'People would give me information because they knew I could easily pass it on to the president. "You need to know this," they would say.'

The first thing John did was to eliminate the traces of Kibaki's predecessor. Moi's official photo came down, replaced by a large one of Kibaki – 'I was very proud of the president' – and a calendar from the Japanese embassy. The civil servants assigned to the department his office replaced – run by a former Moi speechwriter – were sent packing. 'I got rid of all the staff, with the exception of the driver. These guys' loyalties were clear. And in any case, I wanted to get some members of civil society in to lend a hand.' John had an inkling of how institutions and structures can end up insidiously moulding behaviour, rather than the other way round. When the administration assigned him one of Moi's official cars, a dark-blue BMW, he tried driving it around for a day and then returned it, too ill-at-ease to continue.

In came the new team: seven specialists in human rights, governance and the law, picked to roughly reflect Kenya's ethnic diversity. 'Our office was very young. John was the oldest, and he was barely forty,' remembers Lisa Karanja, a barrister and women's rights expert recruited from Human Rights Watch's New York office. With youth, recalls Karanja, came irreverence, absence of hierarchy, and a deliberate adoption of the informal working practices of the non-governmental world from whence so many of the staff hailed. 'We were like an NGO at the heart of government. People would get very shocked, coming into the office, to see John making me a cup of coffee. Here was this powerful man – because he did hold a position of huge influence – and we were calling him "John". The contrast between these new arrivals' breezy directness and other government departments – male-dominated, obsequious and bound by etiquette – was swiftly felt. 'You'd go to meetings with government officials and it would be "your honourable this", "your honourable that" and "all protocols observed" before every speech, all this bowing and scraping,' remembers Karanja. 'I was ticked off at one point for not showing enough respect when I corrected someone who made a legal point I knew was wrong. John had to intervene and say, "Look, she's not here just for decoration. She's not a child."'

The NARC government, in those heady early days, was ready to try something new, opening itself up to its traditional critics. With Moi's departure, thousands of well-qualified Kenyans were returning from the diaspora, ready to put their professional skills at the service of the state. Transparency International, the newly established Kenya National Commission on Human Rights – headed by the outspoken Maina Kiai – civil society groups and human rights organisations were all invited to State House to exchange ideas and offer advice. 'We were sitting down with ministers, and discussing laws. It was unprecedented,' says Mwalimu Mati, who worked at TI-Kenya at the time. 'The only other time there had been anything remotely similar was after independence, when returning students came back to help Jomo Kenyatta's government.'

But there were risks inherent in John's approach, which he would only later learn to appreciate. By surrounding himself with outsiders and failing to cultivate the old guard, he separated himself from the system. When State House colleagues slapped him on the back and joked about 'you and your crazy NGO wallahs' they were underlining a difference that would come to bother them. The policy might insulate John from the sleaze of the previous era, but it also left him dangerously exposed.

There was another failing which John would come to regret: the ambiguity of his job description. 'The first mistake I made was not to sit down and draft precise terms of reference.' He would try repeatedly to pin down his exact role, going to see the president on three separate occasions with ever more explicit definitions, but the initial error could not easily be rectified. His job, as envisaged by Kibaki, was not to formally investigate suspected corruption, nor did he have any powers to prosecute. Those tasks remained with the police force and the attorney general. His role was purely to advise the president – an interesting insight into the extraordinarily centralised nature of power in Kenya – but that duty, in his own view, gave him both the right and the obligation to prod and to poke, to nudge and to pressurise, in any area that seemed to merit attention. 'When people asked about my remit, I would say, "The president asks questions, and I answer them." My job was to pick up the phone and call ten people who could put together a picture allowing me to tell the president what was going on. Essentially, my job was to act as a catalyst.' Such vagueness would be an advantage when he enjoyed the boss's blessing. But it would leave him adrift when things were no longer moving in his direction.

Abandoning lodgings on Lower Kabete Road, in Nairobi's Westlands, John rented a house in the woody suburb of Lavington, close enough to State House to be able to get to work at a moment's notice, however congested the Nairobi traffic. He probably could have claimed a government property – he now had a bodyguard and a small fleet of cars assigned to his office, after all – but he did not want to go down that route: 'If you live in a government house, they own you.' Instead, he found his own place, an elegant villa with a large garden perfect for a pack of bounding, scruffy mongrels he had acquired, specimens of the breed ironically dubbed 'Kikuyu pointers'.

But he never really spent enough time there for it to feel like home. Always a workaholic, John now paused only to sleep and eat. 'We have an eighteen-month window of opportunity before the old, shadowy networks re-establish themselves,' he told reporters, and he was constantly aware of the clock ticking. Here was the chance of a lifetime, the opportunity to put everything he had preached into effective practice, and if it failed, it would not be for want of effort on his part. 'He worked all the time,' says Lisa Karanja. 'It wasn't just late nights, or part of the weekend. He worked *all* the time. So much so that I worried about his health, because he'd had some problems with high blood pressure, and the way he was living didn't seem likely to help.' The dividing line between work and play blurred as John methodically extended an already enormous social circle to include any players with the insights and experience that might help him in the Herculean task of cleaning out Kenya's Augean stables.

Since childhood, John had possessed a talent for bonding with people from different spheres. Thrusting Kenyan yuppies and world-weary Asian lawyers, doddery white leftovers from the days of colonial rule and impassioned activists from Kenya's civil society, lowly taxi drivers and puffed-up permanent secretaries: they might not be able to talk to one another, but they could all, somehow, talk to John. He might not have the hormonal magnetism that allows a man to electrify a crowd, but when it came to the one-on-one encounter, few were more beguiling. Researching this book, I would at first be taken aback and then quietly amused to discover just how many people I spoke to were convinced they enjoyed a special bond with John, sharing unique intimacies and confidences. 'Take it from me, it's you and a thousand others,' I was tempted to tell them. But in a way, they were not fooling themselves. The Big Man had a Big Reach and a Big Appetite. His interest, affection and trust in them were only rarely exaggerated or faked, as far as I could see. It was just that they weren't on exclusive offer. Like a woman of astounding beauty, John was always more loved and desired than he could ever love in return.

It was an instinctive sociability that served his masters well, turning him into a form of national mascot, the acceptable face of the Kibaki regime. John's State House office became a magnet for foreign ambassadors, whose governments picked up its incidental running costs. 'They made it very clear John and our unit were their darling. There wasn't a donor group that came through Kenya that didn't drop in on the office,' remembers Lisa Karanja. For aid officials who had decided to throw their weight behind the Kibaki government's good governance programme and the Western investigators assigned to help it recover Moi's looted assets, a visit served as a refreshing pick-me-up. 'When I came out of my first meeting with John, I felt like I was walking on air,' said one, not normally the type to gush. 'With someone like that in charge of anti-corruption, working for the government, what could possibly go wrong?'

It was impossible to engage with the man and not be won over. Like all charmers, he held up a mirror to those with whom he interacted, and in it they saw a version of themselves. Meeting Kenyan friends his own age he showed his African side: boisterous, loud, irreverent. With the *wazee*, you would think him a son of the soil. When he met with foreign donors, he was more analytical, he knew how to talk the development language of empowerment, benchmarks and governance. 'I never cease to be amazed by the love affair you *wazungus* all seem to have with John Githongo,' Kenyan businessman and columnist Wycliffe Muga once sardonically remarked. But why should he be surprised? John was the perfect dragoman, the go-between who held a pass into two very different universes, Africa and the West. He had lived in both, and could explain Africa's ways to Westerners and the West to Africans. Of course the *wazungus* lapped it up. And State House, looking to donors for renewed funding, was happy to encourage the love affair.

During this period John's family virtually lost sight of him. His parents' hearts might swell with pride every time they saw him on television, but they were worryingly aware that these glimpses afforded their only real insight into what he was doing. 'You wouldn't see him from one month to another,' remembers younger brother Mugo. And when he did turn up, he might as well not have bothered, so seriously did he take the need for professional discretion. 'John is usually a great gossip and storyteller. But at family lunches he would sit and say nothing, just raising one eyebrow,' remembers Ciru, his younger sister. 'It was unbearable. We lost him then, we lost him to the state.'

Old friends still invited John round, but now did so automatically, never expecting him to turn up. His acquaintances, in any case, had long ago coined the term 'to do a Githongo', or 'to be Githongoed', to encapsulate the frustrations that went with being one of John's friends. 'Being Githongoed' meant to be stood up by the Big Man. It meant to be given heartfelt assurances that he would be there, to realise with dawning horror that one had been played for a sucker (again), to sulk a bit, and finally to forgive all when the Big Man resurfaced, so contrite would be his apologies, so rewarding the conversation. 'Githongoing', an area in which all who knew him agreed the otherwise impeccably behaved John regularly performed disgracefully, puzzled me for a while. It wasn't possible, I thought, for a man as rigorous and disciplined as John to confuse his appointments as often as this. Then I realised that his unreliability was in fact the expression of a form of greed: the greed of the intellectual omnivore. When a refreshing new encounter loomed on the horizon, John could not bear to say no. He collected new acquaintances the way others collect stamps, and those joining the collection couldn't help but feel aggrieved on registering that, having once been objects of Githongo fascination, they had been relegated to the category of known quantities, whose exposure to the Big Man would henceforth be strictly rationed.

But John was too busy to worry about such bruised feelings. While overall responsibility for coordinating the anti-graft war rested with the Ministry of Justice, his office would be involved in virtually all of NARC's early efforts to carry out a detailed public tally of Kenya's corruption problem. It was a task only a team as young and absurdly optimistic as John's would embrace with enthusiasm, for it meant probing the roots of a dysfunctional African nation, from the haphazard creation of a British colony to the tortured foundation of an independent state.

4

Mucking out the Augean Stables

'The shocking rot of Nairobi's main market was exposed yesterday when it was revealed that 6,000 rats were killed in last week's cleanup exercise – and an equal number made good their escape. Wakulima Market, through which a majority of Nairobi's three million residents get their food, had not been cleaned for thirty years. So filthy was it that traders who have been at the market daily for decades were shocked to see that below the muck they have been wading through, there was tarmac. More than 750 tonnes of garbage was removed and more than seventy tonnes of fecal waste sucked out of the horror toilets.'

East African Standard, 4 January 2005

In his youth, John had written a Kafkaesque short story about a man who wakes one morning to discover a giant pile of manure has been dumped outside his house. Puzzled, he sets out to establish where it came from and, more importantly, how to shift it. Oddly prescient, the story was a harbinger of John's future task.

Rather than a pile of manure, corruption in Kenya resembled one of the giant rubbish dumps that form over the decades in Nairobi's slums. Below the top layer of garbage, picked over by goats, marabou storks and families of professional scavengers, lies another layer of detritus. And another. With the passage of time the layers, weighed down from above, become stacked like the pastry sheets of a *mille-feuille*, a historical record no archaeologist wants to explore. Each stratum has a slightly different consistency – the garbage trucks brought mostly plastics and cardboard that week, perhaps, less household waste and more factory refuse – but it all smells identical, letting off vast methane sighs as it settles and shifts, composting down to something approaching soil. The sharp stink of chicken droppings, the cabbagy reek of vegetable rot, the dull grey stench of human effluvia blend with the smoke from charcoal fires and the haze of burning diesel to form a pungent aroma – 'Essence of Slum', a parfumier might call it – that clings to shoes and permeates the hair.

As Kenya has modernised, so its sleaze has mutated, a new layer of graft shaped to match each layer of economic restructuring and political reconfiguration. 'In Kenya, corruption doesn't go away with reform, it just migrates,' says Wachira Maina, a constitutional lawyer and analyst. But under all the layers, at the base of the giant mound, lies the same solid bedrock: Kenyans' dislocated notion of themselves. The various forms of graft cannot be separated from the people's vision of existence as a merciless contest, in which only ethnic preference offers hope of survival.

If, in the West, it is impossible to use the word 'tribe' without raising eyebrows, in Kenya much of what takes place becomes incomprehensible if you try stripping ethnicity from the equation. 'A word will stay around as long as there is work for it to do,' said Nigerian writer Chinua Achebe of this taboo noun,[3] and in Kenya, just as in so many African states, 'tribe' is still on active duty. Ask a Kenyan bluntly what tribe he is and he may, briefly, ruffle up and take offence. But the outrage dissolves immediately upon contact with daily life. 'Typical Mukamba, useless with money,' a friend mutters when a newspaper vendor fumbles his change. Another, arriving late at a café, explains: 'I had to straighten up the car because the *askari* was giving me a hard time. Best not to mess with these Maasai.' And when another is fined for parking illegally, he explains: 'I begged with the policeman, but he wouldn't let me off. He was a Kalenjin.'

Any Kenyan can reel off the tags and stereotypes, which capture the categorisation of the country's society. Hard-nosed and thrusting, the Kikuyu are easily identified by their habit of mixing up their 'r's and their 'l's, the cause of much hilarity amongst their compatriots. When an official warns you, 'There may be a ploblem,' a member of civil society denounces 'ligged erections' or an urchin tries to sell you a week-old 'rabradol' puppy, you know you are dealing with either a Kikuyu or his Meru or Embu cousin. Their entrepreneurialism has won them control of the *matatu* trade, and they run most of the capital's kiosks, restaurants and hotels. A Luo, on the other hand, is all show and no substance. His date will be wined and dined, but she'll pick up the tab at the end of the evening. Born with huge egos, the flashiest of dress sense and the gift of the gab, the Luo excel in academia and the media. Luhyas are said to lack ambition, excelling as lowly *shamba* boys, watchmen and cooks. Stumpy, loyal, happy to take orders, Kambas are natural office clerks, soldiers and domestic servants; but watch out for potions, freak accidents and charms under the bed – these are the spell-casters of Kenya. Enticing and provocative, their women dress in eye-wateringly bright colours and often work as barmaids. In contrast, the cold, remote Kalenjin care more about their cows than about their homes. Macho and undomesticated, the proud Samburu and Maasai make for perfect recruits to the ranks of watchmen, wildlife rangers and security guards. And so on . . .

When they speak in this way, Kenyans show, at least, a refreshing honesty. Public discourse is far more hypocritical. In matters ethnic, newspaper and radio station bosses adopt a policy of strict self-censorship. Telling themselves they must play their part in the forging of a young nation state, editors have for decades carefully removed all ethnic identifiers from articles and broadcasts. But it doesn't take long to work out what is really going on, or why one VIP is throwing the taunt of 'tribalist' – Kenya's favourite political insult – in another's face. If a surname isn't enough to accurately 'place' a Kenyan, laborious verbal codes do the trick. A commentator who coyly refers to 'a certain community', or the 'people of the slopes', means the Kikuyu and their kinsfolk from the Mount Kenya foothills. 'People of the milk' indicates the livestock-rearing Kalenjin or Maasai. If he cites 'the people of the lake' or 'those from the west', he means the Luo, whose territory runs alongside Lake Victoria and whose failure to practise circumcision – gateway to adulthood amongst Bantu communities – prompts widespread distrust. The sly euphemisms somehow end up conveying more mutual hostility than a franker vocabulary ever could. Like the ruffled skirts which covered the legs of grand pianos in the Victorian age, they actually draw attention to what they are supposed to conceal: an acute sensitivity to ethnic origin.

The fixation shocks other Africans, who privately whisper at how 'backward' they find Kenya, with its talk of foreskins and its focus on male appendages. 'There's no ideological debate here,' complain incoming diplomats, baffled by a political system in which notions of 'left' or 'right', 'capitalist' or 'socialist', 'radical' or 'conservative' seem irrelevant: 'It's all about tribe.' Directors of foreign NGOs puzzle over the fact that political parties, born and dying with the speed of dragonflies, either don't bother publishing manifestos, or barely know their contents. But who needs a manifesto when a party's only purpose is furthering its tribe's interests? Tribe is the first thing Kenyans need to know about one another, the backdrop against which all subsequent interaction can be interpreted, simultaneously haven, shield and crippling obligation. The obsession is so pervasive, Kenyans struggle to grasp that it may not extend beyond the country's borders. 'So,' commented a Kikuyu taxi driver when he overheard me expressing scepticism about the likelihood of an Obama win in the 2008 US election, 'I see you Westerners have problems with the Luo too.'

Yet, perversely, the strength of these stereotypes is in inverse proportion to their longevity. Rooted in the country's experience as a British colony, Kenya's acute ethnic self-awareness, far from being an expression of 'atavistic tribal tensions', is actually a fairly recent development. While no one would claim that colonialism created the country's tribal distinctions, it certainly ensured that ethnic affiliation became the key criterion determining a citizen's life chances.

* * *

Some time towards the end of the nineteenth century, the story goes, a great Kikuyu medicine man, Mugo Wa Kibiru, woke up trembling, bruised and unable to speak. When he recovered his voice, he issued a terrible prophecy. There would come a time of great hunger, he said, after which strangers resembling little white frogs, wearing clothes that looked like butterfly wings, would arrive bearing magic sticks that killed as no poisoned arrow could. They would bring a giant iron centipede, breathing fire, which would stretch from the big water in the east to the big water in the west, and they would be intent on stripping his people of all they possessed. His people should not fight these strangers. They must treat them with caution and courtesy, the better to learn their ways. The strangers would only depart once they had passed on the secrets of their power.

His prophecy was an uncannily accurate description of the railway that would eventually stretch more than a thousand kilometres from Mombasa on the coast to Kisumu on Lake Victoria. It would never have existed had it not been for William Mackinnon, a Scottish magnate with an evangelical agenda and a romantic appetite for empire, whose imagination was fired by reports brought back by Livingstone and Stanley. The lush kingdom of Buganda, nestling on the shores of Africa's giant freshwater lake in what is today southern Uganda, was blessed with gum, ivory, copra, cotton and coffee. Opening up the hinterland would not only allow its riches to be tapped, it would also, Mackinnon maintained, mean the eradication of the vile Arab slave trade, saving the region for Christian missionaries.

The magnate and his politician friends applied a broad brush when it came to geopolitics, their rough imaginary strokes stretching across half the globe. The recently opened Suez Canal, they argued, held the key to the British Empire's all-important trade with India. If that waterway were to be guaranteed, then the headwaters of the Nile must be secured, and that meant establishing a link between Lake Victoria – source of the Nile – and the coast, controlled by the Sultan of Zanzibar. Above all, a railroad would shore up Britain's position in its long race for regional supremacy with Germany, whose agents lusted after the promised 'new India' just as ardently as Mackinnon.

In 1888, Mackinnon won Queen Victoria's permission to set up a chartered company, the Imperial British East Africa Company (IBEA), to develop regional trade. But constructing the 'Lunatic Line', as the railroad's critics dubbed it, proved beyond IBEA's capacities. By 1895 the company was bankrupt, and Mackinnon handed over responsibility to Whitehall, which announced the establishment of the British East Africa protectorate. Government surveyors set to work, importing hundreds of Indian coolies, thousands of donkeys and camels, and the millions of sleepers required for this monstrous engineering project. The colony that would come to be baptised 'Kenya' was created almost inadvertently, a geographical access route to somewhere seen as far more important.

The railway also played a role in ensuring that Kenya became a settler colony. As construction costs mounted, London became convinced it could only recoup its losses by developing the land alongside the track. '[The railway] is the backbone of the East Africa Protectorate, but a backbone is as useless without a body as a body is without a backbone,' wrote Sir Charles Eliot, the protectorate's new commissioner, in 1901. 'Until a greater effort is made to develop our East African territories, I do not see how we can hope that the Uganda line will repay the cost of its construction.' The proposal seemed uncontroversial, for British officialdom saw few signs of systematic cultivation. Wildlife, in the form of the vast herds of wildebeest, zebra, buffalo and antelope, seemed to outnumber human beings. 'We have in East Africa the rare experience of dealing with a *tabula rasa*,' wrote Eliot, in what must qualify as one of the classic mis-statements of all time, 'an almost untouched and sparsely inhabited country, where we can do as we will.'

Eliot's snap judgement was understandable – a territory the size of France only held around three million Africans at the time, and the activities of both the Kikuyu and the Maasai had recently been curtailed by rinderpest, smallpox and drought. But in fact, much of Kenya's best land was already in use. To the north of the mosquito-plagued stretch of marshy land that would become the city of Nairobi, the well-watered foothills of Mount Kenya were being intensively farmed by the Kikuyu; the nomadic Maasai drove their cattle the length of the Rift Valley; and on the western fringes of this natural cleft Nandi-speaking tribes – later to be rebaptised the Kalenjin – tended crops and livestock. Taming the locals would turn out to require a series of ruthless punitive military expeditions, in which homesteads were set ablaze, herds captured and leaders assassinated.

But the settlers trickled in nonetheless. Fleeing overcrowded Europe, the new tribe dubbed the *wazungu* – 'people on the move' – headed in the main for the Rift Valley's grasslands, which felt more than a little familiar. On a drizzly day, when the chill mists crept stealthily down from the escarpment, they bore a striking resemblance to the rolling heaths of Scotland, a fact that seemed to confirm the settlers in the correctness of their choice. Much has been written about the antics of the dissolute aristocrats who made up the Happy Valley expatriate set. But most of the land-hungry British arrivals in 'Keeenya', as they pronounced it, were from decidedly modest backgrounds, grabbing the chance for a new start. In 1903 there were only around a hundred settlers; by the late 1940s the number had risen to 29,000, boosted by demobilised British soldiers. It would peak at 80,000 in the 1950s. And as the new arrivals marked up their farms, everything began to change for the more than forty local tribes.

Back in Britain, the citizen's right not to have his taxes raised or property confiscated on the whim of a greedy ruler had been recognised since the Magna Carta. But these fundamental principles did not apply to the British Empire's African subjects. A series of regulations passed at the turn of the century decreed that any 'waste and unoccupied land' belonged to the Crown, which could then dispose of it as it wished, usually in the form of 99- and 999-year leases to settlers. In order to force Africans to take paid work on white-owned farms, which were desperately short of labour, the colonial authorities levied first a hut tax and then a poll tax. In the new colony of Kenya, formally declared in 1920, the African citizen was also prevented from competing with white farmers, who alone enjoyed the right to grow tea, coffee, pyrethrum and other crops for export.

The fact that many of the communities the British encountered did not have simple hierarchical structures held up implementation of the new laws only temporarily. The British simply appointed their own chiefs from the ranks of the translators, mercenaries and other 'friendlies' willing to collaborate. It's surely no coincidence that so much power in Kenya today rests in the hands of seventy- and eighty-year-olds who were impressionable youngsters in the years when the draconian colonial regulations made their traumatic impact on African lifestyles. They absorbed vital lessons in how the legal system, the administration and the security forces could be abused to extract labour and resources from an alien land and its resentful people. The first layer on the rubbish tip of Kenyan graft had been deposited.

Inhabitants of pre-colonial Kenya had certainly been aware of their different ethnic languages and customs. But that awareness was a fluid, shifting concept. While sections of the Kikuyu, Maasai and Kamba frequently fought each other over women and cattle, they also traded with one another, intermarried and exploited the same lands, with the pastoralist Maasai, for example, often relying on the agriculturalist Kikuyu to feed their families when drought killed their herds. All that ended with colonialism. Not only did the boundaries drawn by Western powers in the wake of the Berlin conference of 1884–85 slice across the traditional migration routes of communities straddling what had suddenly been delineated as Kenya, Tanganyika and Uganda, the new colonies were themselves subdivided in new, awkward ways. By 1938, Kenya had been partitioned into twenty-four overcrowded native reserves – 'Kamba' for the Kamba people, 'Kikuyu' for the Kikuyu, and so on – and the fertile 'White Highlands' for exclusive European use, where Africans could not own land.

African males were only allowed to travel outside their reserve if they bore the hated *kipande*, an identity card carried around the neck in a copper casket. Introduced to prevent employees from moving to better-paid jobs, the *kipande* corralled Africans inside rigidly defined areas. Wary of anything that could mushroom into a national anti-colonial movement, the authorities banned most political associations; the few allowed were restricted to their founders' ethnic territories. The settlers wanted Africans to act small, think local. It made them so much more manageable.

Registering that white administrators had pigeonholed them, local communities learnt to play the game. Population levels were soaring, thanks to the advent of Western medicine, and the most important asset in a world offering neither pensions, welfare payments nor health insurance – land – would henceforth, they realised, be distributed on a strictly ethnic basis. To those on the reserves, who increasingly viewed their communities as mini-nations in fierce competition with one another, Kenyans from outside were 'foreigners'. The missionaries played their part in this process of self-definition, their translations of the Bible standardising local dialects into formal tribal languages. 'This conversion of negotiable ethnicity into competitive tribalism has been a modern phenomenon,' writes historian John Lonsdale. 'Tribe was not so much inherited as invented.'[4]

The Kalenjin, Daniel arap Moi's ethnic group, represents one such invention. 'Kalenjin' – literally 'I say to you' – was actually the opening line of a series of radio broadcasts used by the colonial administration to muster recruits for the King's African Rifles during the Second World War. It became a label for eight Nilotic communities who shared the Nandi language. Another convenient tag – although this one originated within the community, which saw an overarching tribal identity as lending weight to its dealings with the authorities – was 'Luhya' ('those of the same hearth'), slapped onto twenty subgroups in the 1930s and 1940s. It comes as no surprise to discover that the stereotypes Kenyans apply to one another today, from the fierceness of the Maasai to the supposed domesticity of the Kamba, faithfully reflect the roles the colonial authorities allotted each group: Maasais as mercenaries, Kambas as first porters and then as kitchen workers. Growing up on a white-owned farm in the Rift Valley in the 1940s, the future Nobel Peace Prize-winner Wangari Maathai noticed how the colonial experience reinforced ethnic distinctions. 'Kikuyus worked in the fields, Luos laboured around the homestead as domestic servants, and Kipsigis took care of the livestock and milking,' she records in her autobiography. 'Most of us on the farm rarely met people from other communities, spoke their languages or participated in their cultural practices.'[5]

Two World Wars, in which thousands of Kenyans served, radicalised the colony's African population, challenging this vision of the world. In German East Africa, on the bleak escarpments of Ethiopia and in the jungles of Burma, they saw their white rulers fight and die just like other men. They grasped that the British were mere mortals, their empire beleaguered. The learning experience took place on both sides. 'The younger settlers who had fought in the war with the African had an entirely different outlook on African political advance and the African himself to those who had remained behind,' wrote the pre-independence minister of agriculture Michael Blundell, who led Luo troops to fight the Italians in Ethiopia in 1940. 'The colonial relationship of governing and subject races had been eroded.'[6] Confronted by a range of increasingly belligerent political associations and trade unions calling for a voice in Kenya's administration, London struggled to justify British policy.

The Mau Mau uprising of the 1950s finally exposed the unsustainability of the colonial carve-up. In the run-up to independence in 1963, the regulations that had shaped a sense of separate identity were scrapped, as Africans were granted the right to grow what crops they pleased, to buy land outside the reserves, and to campaign on national issues. But ethnic straitjackets, once tailored, cannot so easily be unstitched. Like so many black leaders of the 1960s, first president Jomo Kenyatta dedicated his energies not to overturning but to inheriting the system left behind by the colonial powers. Only this time it would be his Kikuyu ethnic group, rather than Kenya's departing white tribe, that would benefit from the '*matunda ya uhuru*' – the fruits of independence. While generously helping himself – he taunted former Mau Mau veteran Bildad Kaggia for having so little to show for his liberation war – he made sure his Kikuyu kinsmen got served first when it came to constituency funding, procurement contracts and white-collar jobs in the administration. The fact that no single tribe accounted for more than about a fifth of Kenya's population meant marriages of convenience with at least two other large ethnic groups were always necessary. But priorities were clear. 'My people have the milk in the morning, your tribes the milk in the afternoon,' the president told non-Kikuyu ministers who complained.

When Moi took over on Kenyatta's death in 1978, the approach was perpetuated. Because his Kalenjin ethnic grouping was a smaller, more diverse and less economically powerful group than the Kikuyu, Moi was forced to draw the magic circle a little wider. But Moi's focus remained his own tribesmen, who suddenly found key jobs in the civil service, the army and state-owned companies that had hitherto been closed to them. Ask middle-aged Kenyans today what they consider the root causes of their generation's ethnic wariness, and most point to the education quotas introduced in 1985, which obliged schools to take 85 per cent of their pupils from the local area. The policy was aimed at improving educational standards amongst the Kalenjin, but its impact was to erect even higher walls between communities. Under Kenyatta, at least the tribes learnt mutual tolerance in the playground and classroom. Under Moi, the first time a member of one tribe rubbed up against another was often at university, by which time prejudices had already taken root.

Bullied by Western donors into introducing multi-party politics in 1992, the leader who had done so much to entrench ethnic rivalry presented himself as a national unifier attempting to keep his population's primitive urges in check. 'The multi-party system has split the country into tribal groupings. I am surprised that Western countries believe in the Balkanisation of Africa . . . Tribal roots go much deeper than the shallow flower of democracy.' But if Moi had wilfully reversed cause and effect, he was correct in predicting that the new politics, built on a foundation of rivalry laid by his predecessor and himself, would take ethnic shape. In competitive political systems, argues Paul Collier, parties look for the easiest way to establish their superiority in voters' eyes. Providing services like health, schools and roads is one way of winning approval, but such things are very hard to deliver. Another way is to play the ethnic identity card: 'And that,' says Collier, 'is incredibly easy.'

Analyst Gerard Prunier has christened Kenya's post-independence system of rule a form of 'ethno-elitism'.[7] A pattern of competing ethnic elites, rotating over time, was established which made a mockery of the notion of equal opportunity. This was viewed as a zero-sum game, with one group's gain inevitably entailing another's loss. In Francophone Africa, the approach is captured in one pithy phrase: '*Ote-toi de la, que je m'y mette*' – 'Shift yourself, so I can take your place.' In Anglophone Africa, the expression is cruder, bringing to mind snouts rooting in troughs: 'It's our turn to eat.' Given how unfairly resources had been distributed under one ethnically-biased administration after another, starting with the white settlers, each succeeding regime felt justified in being just as partisan – it was only redressing the balance, after all. The new incumbent was *expected* to behave like some feudal overlord, stuffing the civil service with his tribesmen and sacking those from his predecessor's region. When no one shows magnanimity, generosity dries up across the board.

It's actually possible to quantify the 'Our Turn to Eat' approach in terms of parliamentary seats, ministerial positions and jobs in the state sector, as each regime doled out appointments to those deemed in the fold. According to one study, during the Kenyatta era, the Kikuyu, who accounted for 20.8 per cent of the population, claimed between 28.6 and 31.6 per cent of cabinet seats – far more than their fair share – while the Kalenjin, accounting for 11.5 per cent of the population, held only between 4.8 and 9.6 per cent. With Moi's arrival, the Kikuyu share of cabinet posts fell to just 4 per cent, while the Kalenjin's share soared to 22 per cent. It was a similar story with permanent secretaries, where the Kikuyu went from 37.5 per cent under Kenyatta to 8.7 per cent under Moi, while the Kalenjin went from 4.3 per cent to 34.8 per cent.[8]

In theory, of course, a particular ethnic group could hold the lion's share of key government jobs without it distorting national policy. In fact, the entire arrangement was premised on the pork-barrel principle. Hoeing their Central Province plots in bare feet and ragged hand-me-downs, a minister's constituents might feel they had little, individually, to show for their community's pole position. The top men stood at the apex of frustratingly inefficient pyramids of dispersal. But what was the alternative? 'The grassroots perception is, if we elect a member of our elite, he can at least talk to the elites of the other tribes,' says Haroun Ndubi, a human rights campaigner. 'People will say: "This is someone who can speak English with the others."' And if a local hero consistently failed to pass at least a fraction of what came his way along the chain, he could expect to be unceremoniously dumped come the next election.

The difference being on the right side made was illustrated when the ministry for roads and public works published estimates for spending on road-building in July 2006. Regions whose MPs formed part of Kibaki's inner circle got far more than was allocated to areas whose leaders were in opposition. Once Nairobi and the tourist hub represented by the Maasai Mara were excluded, allocations to the home constituencies of vocal government critics were nearly 320 times less generous than those to constituencies of trusted presidential aides.[9] The parliamentarians made some barbed remarks when this extraordinary gap was exposed, but passed the road budget without amendment. This, they knew, was the way the game was played.

Where does each individual draw the limits of his or her compassion, beyond which duties of kindness, generosity and personal obligation no longer apply? I was raised in a household where my parents drew them in totally different places, according to their very different characters and backgrounds.

As an Italian, my mother grew up in a country whose government had given birth to Fascism, formed a discreditable pact with Hitler, and launched itself on a series of unnecessary wars which left Italy occupied and battle-scarred. There then followed a seemingly endless series of short-lived, sleaze-ridden administrations. The experience left her utterly cynical about officialdom. Although she dutifully voted in every election, the malevolence of the system was taken for granted, and she would happily have lied and cheated in any encounter with the state had she believed she could get away with it. But no one worked harder for her fellow man, for in the place of the state she maintained her own support network. An instinctive practitioner of what sociologists call 'the economics of affection', my mother had a circle of compassion drawn to include a collection of needy and lonely acquaintances. She visited their council flats bearing cakes, sent amusing press cuttings to their prison cells, queued at the gates of their psychiatric hospitals. Hers was a world of one-on-one interactions, in which obligations, duties, morality itself, took strictly personal form, and were no less onerous for it. The glow she radiated was life-enhancing, but its light only stretched so far, and beyond lay utter darkness. Protecting one's own was vital, for life had taught her that the world outside would show no mercy. She was not alone in her ability to get things done without the state's involvement. '*Il mio sistema*' Italians call it: 'my system'. Italy is, after all, the birthplace of the Mafia, the ultimate of personal '*sisteme*', and my mother's mindset was instinctively *mafioso*.

My father, in contrast, was typical of a certain sort of law-abiding, diffident Englishman for whom a set of impartial, lucid rules represented civilisation at its most advanced. He was raised in a country which pluckily held out against the Germans during the Second World War and then set up the National Health Service in which he spent his career, and his trust in the essential decency of his duly elected representatives was so profound that he was shocked to the core by British perfidy during the Suez crisis, and believed Tony Blair when he said there were weapons of mass destruction in Iraq. When, as an eleven-year-old schoolgirl, I mentioned – with a certain pride – that I usually managed to get home without paying my bus fare, he explained disapprovingly that if everyone behaved that way, London Transport would grind to a halt. Remove the civic ethos, and anarchy descended. A logical man, he saw this as the only practical way of running a complex society. It also, conveniently for an Englishman awkward with personal intimacy, enabled him to engage with his fellow man at a completely impersonal level. Not for him my mother's instinctive charm, the immediate eye contact, the hand on arm. He felt no obligation to provide for nieces and nephews, and had a cousin come up for a job before one of the many appointment boards on which he sat, he would have immediately excused himself. Nothing could be more repugnant to him than asking a friend to bend the rules as a personal favour. What need was there for a rival, alternative *sistema*, if the existing arrangement of rights and duties already delivered?

My father's world view was typically northern European. My mother's characteristically Mediterranean approach would have made perfect sense to any Kenyan. In an 'us-against-the-rest' universe, the put-upon pine to belong to a form of Masonic lodge whose advantages are labelled 'Members Only'. In the industrialised world, that 'us' is usually defined by class, religion, or profession. In Kenya, it was inevitably defined by tribe.

Western analysts have remarked on Africans' 'astonishing ambivalence' towards corruption,[10] but it is not so surprising. Under the colonial occupiers and the breed of 'black *wazungus*' who replaced them, the citizen had learnt to expect little from his government but harassment and extortion. 'Anyone who followed the straight path died a poor man,' a community worker in Kisumu once told me. 'So Kenyans had no option but to glorify corruption.' In a 2001 survey, Transparency International found that the average urbanite Kenyan paid sixteen bribes a month,[11] mostly to the police or the ministry of public works, to secure services they should have received for free. Added together, *kitu kidogo* – supposedly 'petty' corruption – accounted for a crippling 31.4 per cent of a household's income. Those paying out no doubt saw themselves as innocent victims of oppressive officialdom. But while chafing at the need to grease palms, ordinary Kenyans were also playing the system with verve. Which of them could put their hand on their heart and swear that they had never relied on a 'brother' for a bargain, a professional recommendation or a job? Who had never helped a distant 'cousin' from upcountry jump a queue or win special access? Aware of their own complicity, they hesitated to point an accusing finger.

Moral values can become strangely inverted in a harsh environment. 'In Nairobi, around 50 per cent of the population is either unemployed or underemployed – they're selling shoelaces or picking up rubbish, not earning enough to survive. But this country doesn't have soup kitchens, and you don't see hordes sleeping rough,' says Professor Terry Ryan, a veteran Kenyan economist. 'That's because a senior civil servant or CEO typically picks up the school fees and hospital bills of roughly fifty of his kinsmen, while a headmaster or low-ranking civil servant will be supporting twenty-nine members of the extended family in one way or another.' Propping up such vast networks made bending the rules virtually obligatory. The man who abided by the rules and took home no more than his salary seemed to his relatives a creep; the employee who fiddled the books and paid for his aunt's funeral, his niece's education and his father's hernia operation a hero.

In a poor country, ethnic marginalisation does more than blight life chances. It can actually kill. A 1998 survey found that Kalenjin children were 50 per cent less likely to die before the age of five than those of other tribes, despite the fact that most lived in rural areas, where life is generally tougher.[12] The statistic makes perfect sense. Under Moi, Kalenjin areas benefited from better investment in clinics, schools and roads. A worried Kalenjin mother would head for a well-stocked nearby clinic, child in her arms, along a smoothly tarmacked road. Her non-Kalenjin equivalent was likely to be tossed for hours in the back of a *matatu* struggling along a rutted track, only to eventually reach a clinic with nothing but aspirin on its shelves and watch her child die.

Researching this book, I repeatedly asked Kenyans for examples of how ethnic favouritism had personally affected them. 'Oh, *every* Kenyan has a story like that,' I was always told. Yet few volunteered details. It was easy to guess why. If they had lost out because of tribal patronage, they risked looking like whiners; if they had benefited, they'd be admitting to collaborating in a system that fostered incompetence.

I'd seen one example myself, at a Kenyan newspaper where I briefly worked as a subeditor. The *East African Standard* was being revamped after many years in the doldrums. The details of its ownership had always been kept deliberately murky, but the fact that the Moi family quietly pulled the levers was widely known, and had alienated readers, while management's habit of giving jobs to barely literate Kalenjins was blamed for a general collapse in standards. Now a new chief executive was poaching talent from rivals, with promises of an imminent takeover by a South African company. After my first few weeks at the paper, I went for lunch with one of the senior writers.

'So, what do you think of the staff?' he asked.

I ran through my various colleagues. Some had better training than others, some were more enterprising, but the goodwill was undoubtedly there. With one exception. The man in question, I said, turned up late or not at all, lounged at his desk playing music while the others hammered at their keyboards, and was often rude to his fellow workers. Robert – let us call him – was one of those rare, dangerous subeditors who could take a perfectly decent story and insert fresh mistakes. When I'd pointed one of these out, he'd given me a look of such astonished contempt that I'd realised criticism was something he rarely heard. In a surprisingly short space of time I'd come to detest him, and it was clear to me that many staff felt the same, although they were strangely mute in his presence. The man should obviously be fired.

The journalist gave me a long look, enjoying his moment.

'You'll be interested to hear that I expect that individual to either take my job very shortly, or be made editor.'

Robert, he explained, was a close relative of one of the newspaper's top executives. Both men were Kalenjins. No matter how incompetent or unpleasant, Robert knew his career was assured – hence his arrogance and his co-workers' resentful silence.

'Good Lord,' I said. It was so crude I could barely believe it. 'You know, where I come from, the boss's son often works twice as hard to make sure people don't accuse him of exactly this form of nepotism.'

He shrugged. 'Not here.'

'How about sending him on a training course so that, at the very least, he learns his trade?'

'Oh, that's been done. Few people at the newspaper have received more training. He's even gone on one of those journalism courses in the UK. He never gets any better. It's a question of attitude.'

Having written about ethnic patronage for years, this was the first time I'd seen up close its insidious impact on a workplace. Since that lunch, many of the people we discussed that day – including my lunch companion and the chief executive who had dangled the hope of a South African takeover – have left the newspaper, which remains in Moi's control. They had been mis-sold the notion of a meritocratic, non-tribal, politically independent company, and with that promise went much of the incentive for staying. Robert, in contrast, has been promoted, just as predicted.

That was my story. But I wanted to hear someone else's.

Eventually I found him. His name was Hussein Were. He was forty-two when we met, and his boyishly unlined face jarred with the methodical manner of a much older man. Deliberate and self-contained, he spoke at perfect dictation speed – no rushing or interruptions permitted – and his sentences, peppered with 'albeit's and 'pertaining to's, were redolent of the legalistic world of depositions and affidavits, in which people pause before speaking and are careful to say what they mean.

His first job, he told me, had been with a firm of quantity surveyors, where he spent more than ten years. The boss was a Kamba and a Christian, Were a Muslim and a Luhya, but that didn't stop them working well together. So well, in fact, that when Were tendered his resignation, explaining that he had won a scholarship for a Masters at the University of Nairobi, his boss persuaded him to stay on, juggling his day job with his studies. But when Were returned to full-time professional life, he noticed that things were changing. The company was expanding, and every new arrival, he registered with quiet dismay, was a Kamba. 'The assistant was Kamba, the secretary was Kamba, the receptionist was Kamba. It was becoming a single-ethnic organisation.' His relations with these staff were cordial. They shared lunches, knew each other's families. But Were began feeling excluded in subtle ways. 'In those situations, people begin to segregate into groups. They regard you as different and don't want to share certain things. They set up informal networks, channels inside the office.' He did not understand the language in which the others communicated, and as a Muslim he would not be included in any Friday-evening trip to a local bar.

Were gritted his teeth. He had hoped for better – 'Maybe I'm naïve' – but he felt no real surprise. 'I had come of age learning about the working environment in this country. I knew Kenya was full of one-ethnic companies. I thought, "I'll live with it."' His ambitions remained high. After ten years in the job he had every reason to expect to be made partner. Then professional rivalry began to undermine his reputation for efficiency. 'If I was registering certain successes, my colleagues wouldn't want them to reach the boss. But negative things would immediately be brought to his attention.' Were, who had once been his boss's second-in-command, noticed that key information was now passing him by. He was being written out of the script. 'Colleagues would mention things that concerned me directly that they had been discussing separately with my boss, chats which were probably taking place during visits to construction sites.' At that stage, Were resigned. 'I saw the whole thing was untenable.'

He didn't bother to explain why he was going. 'I never raised it directly with my boss, because I realised he was encouraging it. I just said I needed to progress my career.' Like many Kenyans caught in such circumstances, he expresses not anger, but resignation at what he knows to be a commonplace experience. 'There are lots of people in this country who have never sat a job interview or even know what one is. They have been whisked by their tribespeople from school to job. I believe in fighting my own way.' Friends tell him his problem was not being 'anchored' by a network of friendships and family relationships that would have made it impossible to 'detach' him from his place of work. But he has no intention of developing these limpet-like muscles. At the consultancy he has now set up, he's proud of the fact that not a single one of his current projects comes from a fellow Luhya. 'There are people who feel like me, who do not subscribe to that kind of thinking,' he insists. 'I wouldn't pack a company with my people.'

* * *

Were's experience, and that of my colleagues at the *Standard*, was the most benign manifestation of the 'Our Turn to Eat' culture. Its other forms were much uglier, and their impact far more damaging. So few Kenyans identified with any overarching national project, their leaders felt free to loot state coffers, camouflaging crude personal enrichment in the prettifying colours of tribal solidarity.

Decade by decade, practices that had flourished under the colonial administration – itself no stranger to high-profile corruption scandals – were fine-tuned and pushed to ever more outlandish lengths. What they all shared were a reliance on the political access and inside knowledge enjoyed by either a minister, an MP, a civil servant or a councillor, and their target: the public funds and national assets on which every Kenyan citizen depended for education, health and the other basic necessities for a decent life.

The command economy of the post-independence years made self-enrichment for the well-connected a fairly simple matter. What could be easier for a minister than to slap an import quota on a key commodity, wait for the street price to soar, and then dump tonnes of the stuff, thoughtfully stockpiled ahead of time by one of his companies, on the market? A 1970–71 parliamentary commission helpfully authorised government employees to run their own businesses while holding down civil service jobs ('straddling', as it was called), a ruling its chairman later justified on the grounds that there was no point banning an activity that would persist whatever the law decreed.[13] A post in a state-run utility or corporation, which could hike prices ever upwards thanks to its monopoly position, offered untold profit-taking opportunities. Similarly, who was better placed to benefit from foreign exchange controls which created a yawning gap between black market and official rates than an insider with excellent banking and Treasury contacts?

The structural adjustment programmes pushed on Africa by the World Bank and the International Monetary Fund in the 1980s, which loosened the Kenyan government's stranglehold by making aid conditional on privatising bloated parastatals, dropping currency controls and opening markets to international trade, complicated things, but the 'eaters' quickly vaulted that hurdle. The privatisation process itself, it turned out, provided all kinds of openings for the entrepreneurial fraudster, including ruthless asset-stripping. It was funny how often the politically-connected banks in which state corporations chose to deposit their proceeds collapsed, swallowing up public funds as they expired. And so many other routes remained open. Import goods duty-free as famine relief, or claim they are in transit, then sell them locally, undercutting the competition. Take out a state loan you never intend to repay. Bid for a government tender your contacts at the ministry tell you is about to come up, then get them to ensure that your ridiculously inflated offer is the one approved. It doesn't matter if your firm can't deliver: the invoice will join Kenya's huge stock of 'pending bills', carried over from one government to another, and eventually settled with the issue of trade-able treasury bonds, no questions asked.[14]

By the early 1990s, Western executives flying in with plans to invest in Kenya quickly realised that their companies would never thrive in the country's supposedly free-market environment unless a slice of equity was discreetly handed over to a firm owned by a Moi relative, trusted henchman or favoured minister. Frank Vogl, who runs a communications firm in Washington, caught the flavour when he was approached to set up a presidential press unit. Summoned by Kenya's finance minister to discuss the idea, he flew to Nairobi and went to the minister's offices. 'It was so full I could barely squeeze in the door. The entire reception area was jammed with about twenty or thirty people, who were all trying to reach the secretary sitting at reception. I finally managed to catch her attention and said: "I have a 10 o'clock appointment with the minister." "So does everyone else," she said. "You'll have to wait your turn." These were all businessmen waiting to have their one-on-ones with the minister – and you can imagine just what was going on during those conversations. It was no longer a secret by then: if you wanted to do business in Kenya, you had to do a deal with the top man concerned.'

And spanning every regime was land-grabbing, which pushed so many African buttons. Swathes of supposedly protected game parks, plots already owned by state-run corporations and municipal bodies, prime sites on the coast, chunks of gazetted virgin forest lusted after by timber merchants, were snatched, fenced off and sold on again. The practice was so widespread that even the leaders of Kenya's churches, mosques and temples – society's supposed moral arbiters – joined in. The grabbers did not hesitate to seize plots set aside for national monuments or already used as cemeteries, simply throwing the bodies onto the street. The phenomenon peaked before elections, as the president of the day thanked his cronies in advance for their support. Inquiries would reveal some 300,000 hectares of prime land to have been seized since independence, with only 1.7 per cent of the original 3 per cent of national territory gazetted as forest remaining – jeopardising a thirsty nation's very water table.

But 'eating' surely touched its nadir with the Goldenberg scandal, the Moi presidency's crowning disgrace. Dreamt up by Kamlesh Pattni, a Kenyan Asian with a lick of glossy black hair and the over-confidence of a twenty-six-year-old millionaire, this three-year scheme was once again a reflection of its times.[15] Launched in 1991, it tapped into the government's hunger for foreign exchange, threatened by aid cuts from Western donors determined to see multi-party elections in Kenya. Pattni's firm, Goldenberg International Ltd, started by claiming – under a government compensation scheme meant to encourage trade – for exports of gold and diamonds Kenya did not produce and the firm never actually carried out. Approved by Central Bank staff, Pattni's fraudulent export forms – the infamous 'CD3's – only marked the start of this multi-layered scam. Setting up his own bank, he used the leverage granted by his finance ministry contacts to mop up available foreign exchange under a pre-shipment finance scheme. He bought billions of shillings in treasury bills on credit and cashed them in as though they had been paid for, and borrowed money from a range of complicit 'political banks' to place on overnight deposit.

The various schemes not only enriched senior officials, they provided slush funds for what the ruling party knew would be fiercely contested elections. Pattni ploughed his profits into the construction of the Grand Regency, a five-star hotel in central Nairobi as gilded and ornate as Cleopatra's boudoir. The ordinary Kenyan, for his part, lost anywhere between $600 million and $4 billion as his country's foreign exchange reserves, rather than being boosted, were systematically hoovered up by the well-connected. Goldenberg pushed the country's inflation into double digits, caused the collapse of the Kenya shilling and a credit squeeze so severe it led to business closures and mass sackings, and left the government unable to pay for oil imports and basic health and education. The resulting recession was still being felt fifteen years later.

Goldenberg captured the very essence of Kenyan corruption. For if only a tiny elite got obscenely rich on the back of it, the sleek Pattni carefully shored up his enterprise with a liberal distribution of gifts: a form of insurance. The astonishing extent of wider Kenyan society's complicity would only be exposed in 2004 when investigators published a list of those alleged to have benefited from Pattni's largesse. Gado, the *Nation*'s brilliant cartoonist, captured the moment with one of his sketches. 'Anybody who has not received Goldenberg money, please raise your hand,' runs the caption. Below, a variegated cross-section of Kenyan society stares at the reader, boggle-eyed, uncomfortable, shifty: a bewigged lawyer, a Muslim preacher, a portly *mzungu*, a stout matron, a notebook-wielding journalist, a uniformed nurse, a scruffy panhandler. No one moves. All, at one point, have benefited from Goldenberg. The 'list of shame', as it was dubbed, ran to 1,115 entries.

5

Dazzled by the Light

'Africans are the most subservient people on earth when faced with force, intimidation, power. Africa, all said and done, is a place where we grovel before leaders.'

JOHN GITHONGO, *Executive* magazine, 1994

Working alongside the director of public prosecutions and a brand-new ministry of justice – an institution phased out under Moi – John Githongo had the job of digging down through this purulent history, sorting through the layers of sleaze.

The judiciary, which had become stuffed over the years with bribable magistrates ready to do Moi's bidding, must be purged: scores would eventually be publicly denounced, dismissed or encouraged to retire. Ministry departments needed to be cleansed of a generation of bent senior procurement officers who had for decades used public procurement as a source of illicit wealth, stealing, one study estimated, \$6.4 billion between 1991 and 1997.[16] An inquiry, the Bosire Commission, was launched to probe the Goldenberg scandal. Another, the Ndung'u Commission, probed the land-grabbing phenomenon. Yet another was established to investigate the scandal of pending bills. In a grand gesture of good faith, Kenya also became the first country in the world to ratify the UN Convention against Corruption.

Then there were the two pieces of legislation Kibaki had announced on the lawns of State House soon after his inauguration:

the Public Officer Ethics Act, which spelt out a code of conduct for public officers and obliged them to declare their wealth; and the Anti-Corruption and Economic Crimes Act, which created the Kenya Anti-Corruption Commission (KACC), a doughty successor to the anti-corruption authority set up but rapidly neutered under Moi.

John helped ensure that the directorship of the new institution, which he eventually hoped to see given prosecutorial powers, went to Justice Aaron Ringera, whom he had befriended during his time at TI-Kenya on a long-haul flight to a World Bank meeting. Convinced that this former solicitor general was the perfect candidate for the job, he went in person to lobby the various political party leaders – not all of whom shared his enthusiasm for Ringera – to support the appointment. 'I put my reputation on the line, without hesitation or equivocation. I had complete faith in Ringera.' John was also partly responsible for the KACC director being granted one of Kenya's most generous civil service pay awards. The bigger the salary, the easier it would be for the holder of this key institution to resist temptation, he told the sceptics.

In NARC's flurry of law-making, one thing, however, was made clear. These inquiries would not go to the very top of the chain. Moi's lieutenants might be vulnerable to prosecution, but the former president himself would remain beyond pursuit. The new administration justified this stance on the grounds that ordinary Kenyans, grateful for Moi's tactful withdrawal from the political scene, would be revolted by the sight of a venerable elder being hounded through the courts. It was an argument John endorsed. He should have been more alert to the gesture's underlying message. Even in the new-look, squeaky-clean, corruption-phobic Kenya, the really big players could expect to get off scot free, while the smaller fry would be held to account.

As he put in his endless working days, friends from the old days noticed with concern that John, originally taken on as a consultant, now spoke in terms of 'we' when referring to State House. It was 'our government', 'our administration', and when cynics expressed scepticism, he grew annoyed, for it meant doubting John himself. Having decided that NARC represented Kenya's best chance to tackle a deep-rooted blight, he had deliberately failed to install a safety net. Some saw this as a step further than was wise, or was warranted by his job description. 'He was using the language of government, when he should have seen himself as someone who had been seconded to government,' says anti-corruption campaigner Mwalimu Mati. 'He should have retained an intellectual distance, seen himself as an adviser, a specialist.'

Others viewed it as typical of a man who had to believe passionately in his allotted task to function at all. 'He went into it with a lot more idealism than I thought warranted. But John is a conviction person, it's a personality type,' says David Ndii. 'With him, it's all about the heart. When John trusted someone, he did it completely. And when he was disappointed, he flipped completely. He has this pendulum thing.' Beguiled by the sheer physical solidity of the man, his elders missed this emotional volatility. It made John a far more unpredictable player than those who had appointed him realised. 'He probably didn't have the right character for the job,' says Ndii. 'Government is all about perseverance. John was disposed to the melodramatic.' The balked romantic can prove surprisingly vindictive, turning avenging angel where others might simply withdraw into a sulk.

As a journalist, John had railed against two weaknesses he saw as intrinsic to his continent's predicament: the extraordinary deference African societies traditionally show their elders, and their meek passivity when confronted by rulers ready to use violence to remain at the helm. Moi, famously, had instructed his ministers to 'sing like parrots'. 'You ought to sing the song I sing,' the president had told his cabinet. 'If I put a full stop, you should also put a full stop. That way the country will move forward.' The crudeness of the order, the exhortation to abandon all critical thought, argued John, exposed a humiliating respect for power for its own sake. Yet now that he was within the citadel, both insights momentarily eluded him. 'There was a reverential tone in John's voice when he talked about Kibaki,' remembers Rasna Warah, a columnist for the *Nation* and an old acquaintance. 'It would be "the president thinks this", "the president wants that", never just "Kibaki". It was a tone of total awe, as though the man had become a living saint.'

If he had fallen prey to Strong Man syndrome, John was not the only smitten one. Bubbling with hope, the entire country needed, for a moment in history, to forget what it knew about Kibaki and his chums. Nations must indulge in periods of selective amnesia if they are ever to progress. History suggests that sclerotic systems are not transformed by untainted outsiders, but by those within, and usually by those who have been within the system so long they are associated with its worst abuses, rising thereby to the positions of power that make it possible to bring about change. Mikhail Gorbachev was such a figure in the Soviet Union – a seemingly loyal party stalwart who turned radical once he had the means to see his novel vision through.

On the surface, there was little reason to view Kibaki, who had played the Kenyan system to the hilt as both vice president and finance minister, as a likely champion of reform. The first African to graduate from the London School of Economics, a former lecturer at what became Uganda's respected Makerere University, one of the drafters of independent Kenya's constitution, Kibaki was routinely described as 'brilliant'. But his glory days lay firmly behind him. He was remembered as the man who had seconded a 1982 motion making Kenya a *de jure* one-party state and loftily dismissed opposition attempts to topple KANU as 'trying to cut down a giant fig tree with a razor blade'. Having swallowed one political humiliation after another under Moi, his preference for the unconfrontational role of Mr Nice Guy had won him the scornful sobriquet of 'General Coward' from political rivals, who quipped that Kibaki had never seen a fence he couldn't sit on. Well-heeled, well-oiled, Kibaki's image as a prosperous has-been was so entrenched by the mid-1990s that it never occurred to Western journalists like myself to request an interview. Why bother? The nominal head of the opposition was reported to be a sozzled regular at the Muthaiga Golf Club, interested in little more than the size of his handicap. While he regularly drove to parliament, he rarely performed inside the chamber, preferring, it was said, a long snooze at his desk. Yet suddenly this deeply disappointing politician was recast in the role of national saviour by a coterie that, believing it held the moral high ground, thought nothing was now impossible. 'Go home, tend to your goats and watch us govern this country,' justice minister Kiraitu Murungi told Moi, courting hubris with every patronising word.

'They got lost in their own rhetoric,' says Ndii, with a shrug. 'Because they had the instruments of state, they thought they could change the world. It wasn't just John, all of them thought they were going to fix everything. Me, I was not a believer.' Mwalimu Mati also shakes his head over what looks, in retrospect, like the most bizarre of collective delusions. 'It was a type of mass hallucination. People went a bit crazy. No one stopped to consider how Kibaki had made his own fortune. We should be suspicious of finance ministers, especially from the past.'

It may have been a case of the ultimate idealist meeting the ultimate pragmatist, but John did not recognise the gulf in perspectives. Bonding with Kibaki came disconcertingly easily. A politician with none of Moi's instinctive understanding for the ordinary *wananchi*, Kibaki was an unrepentant intellectual snob. Whereas Moi, the former headmaster, was regarded as a leader who 'knew how to talk to Kenyans with mud between their toes', Kibaki was more likely to hail them as '*pumbavu*' – fools. He recognised and respected the rigorous quality of thought in the young man, who had strayed into State House at more or less the same age Kibaki himself had ventured into politics. There was also a certain inbuilt familiarity to the relationship. John's accountant father had campaigned on behalf of Kibaki's Democratic Party, and while the Kibaki and Githongo families were not exactly intimate, their children had gone to the same schools, they shared the same faith, they belonged to the same patrician milieu.

In any case, affability came naturally to Kibaki, who possessed none of Moi's gruff abrasiveness. While other men commanded loyalty through the commanding magnetism of their personalities,

Kibaki's style was one of diffuse, woolly bonhomie. He had always shrunk from making enemies, the head-on collision. 'He's a very unstuffy guy, very laid back and easy to shoot the breeze with,' John remembers. The two regularly breakfasted together, and there were also many dinners, just the two of them tête-à-tête. Kibaki felt relaxed enough in John's company to sit with him in the presidential bedroom, discussing politics, the price of oil, world affairs – never anything personal. In John's slightly star-struck eyes – who, after all, could spend quite so much time near the nation's most important man without feeling a little giddy? – the president came to assume the role of alternative father figure, favourite uncle. If John used the respectful '*Mzee*' (Elder) when addressing the president, Kibaki addressed his anti-corruption chief as '*Kijana*' – 'young man', a term that almost always comes tinged with paternal affection. 'I used to think that relationship was very special. I had a huge amount of affection for Kibaki. Then I realised Kibaki was like that with everyone.' Looking back, John would come to realise that he had allowed himself – as the overly cerebral often do – to be beguiled as much by a symbol as an individual. 'At that time, everyone was dancing. Everyone was *right* to dance.' Encapsulating the hope of a jubilant post-Moi nation, what Kibaki represented was more important than who he actually was.

John had the goodwill of the head of state, the envy of many veteran political players, his own staff and budget. It seemed, on the face of it, a great set-up from which to take on the forces of darkness. But within weeks of Kibaki's inauguration, the evil genie Moi deposited in State House snickered and lashed out, delivering a blow so devastating, so sudden, that the presidency, it could be argued, never recovered. Kibaki's presidency was delivered premature, shrivelling before it had a decent chance to take its first real breaths. A crippled and maimed thing, it would be too worried about its own survival to care overmuch about anything else.

* * *

The first Kenyans heard of it was an announcement, in late January 2003, that the president had been admitted to Nairobi Hospital to have a blood clot – after-effect of his car accident – removed from his leg. Kibaki would continue to carry out his official functions from hospital, his personal doctor Dan Gikonyo assured the public, as long as he did not get overstressed. He suffered from high blood pressure and had been advised, amongst other things, not to wave his arms around. The statement failed to reassure. 'I don't want to cause alarm but I am worried about our president's health,' a perceptive Kenyan blogger wrote in February, noting that Kibaki had not addressed the nation for a month, remaining silent even when a minister was killed in an air crash. 'I have this nagging feeling that State House is not telling all.' The blogger quoted eyewitness accounts of an incoherent president checking out of hospital and embarking on a strange two-hour meet-the-people drive around Nairobi. 'Something is wrong, something is terribly wrong,' he fretted.

Kibaki had, in fact, been felled by a stroke. Any debate about how many terms he hoped to serve was suddenly rendered irrelevant – would he even see one through to the end? When John Githongo went to visit the Old Man in hospital, he was shocked. Whatever criticisms had been voiced of Kibaki in the past, everyone had agreed on his extraordinary intellectual acuity. Now John found him watching television cartoons. He never mentioned his new concern to friends, but the worrying vision of Kenya's top statesman happily transfixed by children's programming lingered in his mind: 'You never completely recover from a stroke like that.' Once Kibaki checked out of hospital, John started briefing him both orally and in writing, so concerned had he become over his boss's ability to retain information.

Journalists who covered NARC's 2002 election campaign say there have been two Kibakis: the pre-stroke Kibaki, engaged, focused, acute; and the post-stroke Kibaki, vague, distracted, struggling to maintain a coherent chain of thought. From a man in command he had become a man going through the motions, as if in a dream. The British high commissioner, Edward Clay, immediately noticed a change. Just as Britain, traditionally a major donor, was hoping to re-engage with Kenya, it became impossible to win an audience with the president. Development minister Clare Short left the country without seeing the head of state. And Clay noticed that Kibaki struggled during his regular meetings with the diplomatic corps. 'He had a genuine problem carrying on a train of thought from one meeting to another, particularly if there wasn't a witness. Some days were better than others. I didn't think he was himself again until early 2004.'

It was noticeable that when Kibaki was delivering a speech he no longer extemporised or made eye contact with his public, keeping his eyes glued to the autocue. He knew that if he lifted his gaze he might never find his place again. There were reports of him sleeping through cabinet meetings, of aides having to repeatedly brief him on the same subject. At an investors' meeting I attended in London two and a half years after his collapse, by which time many were remarking on the extent of his recovery, Kibaki still gave the impression – characteristic of stroke victims – of being a little tipsy. His delivery was slightly slurred, his enunciation ponderous, and when answering questions he meandered and contradicted himself. The entire audience seemed to be willing him on, praying he would make it through to the end without some monstrous *faux pas*. Like the latter-day Ronald Reagan in the grips of early Alzheimer's, he came across as an urbane, delightfully charming old duffer, but not a man anyone would want running a country.

Confronted by a calamity no one had anticipated so early on, Kibaki's closest aides reeled and then rallied. If the Old Man was temporarily incapacitated, then they would have to run the country until he regained his faculties, just as the Kremlin's stalwarts had done whenever their geriatric Soviet leaders turned senile. The kernel of this group consisted of Chris Murungaru, the burly former pharmacist appointed minister for internal security; David Mwiraria, finance minister and Kibaki's longtime confidant; Kiraitu Murungi, justice minister; State House comptroller Mateere Keriri; and personal assistant Alfred Getonga. The one factor all these players had in common was their ethnicity – they were all either Kikuyu, like Kibaki, or members of the closely related Embu and Meru tribes, who the Kikuyu regard as cousins. In naming his cabinet, Kibaki had presented himself as a leader of national unity, careful to distribute all but the key ministries across the ethnic spectrum. But in his hour of need, like any sick man, he reached for what was familiar and safe, and that meant sticking with the tribe. The popular press, noticing the trend, soon coined a phrase for this circle, the real power behind the throne. 'The Mount Kenya Mafia', it called them, a reference to the mountain that dominates Central Province. The phrase was to prove more apposite than anyone could have guessed at the time.

The group's influence was swiftly felt in a vital area. A new constitution had been one of the key promises NARC had made to an electorate exasperated at the way in which Kenya's colonial-era document had been repeatedly amended to place ever greater power in the president's hands. Kibaki had also, it emerged, signed a memorandum of understanding with his NARC partners promising, amongst other things, that fiery Luo leader Raila Odinga would be given the post of executive prime minister under a future dispensation. Incapacitated by his car accident, Kibaki had depended on Raila to do his heavy lifting during the election campaign, and the younger man had done so indefatigably. The prime minister's post was to have been his reward. It was a promise that implied a radical trimming of powers in favour of a tribe that Kibaki's Kikuyu community had, since the days of Jomo Kenyatta and Raila's late father Jaramogi Oginga Odinga, regarded as its greatest rival. After decades of marginalisation, during which they had seen their leaders assassinated, jailed and exiled, the thwarted Luos were itching to come in from the cold.

But now, with Kibaki looking like the weak old man he was, all promises were off. The Mount Kenya Mafia felt too vulnerable for magnanimity. The very same men who had, as members of the opposition, tirelessly denounced a document that skewed the playing field in Moi's favour, suddenly found there was much to be said for this tilted arrangement. A national conference convened to hammer out the modern arrangement Kenya needed became gridlocked, as Kibaki's key ministers proposed changes that would, if anything, concentrate even more power in their man's hands. The Kibaki delegation would eventually storm out of the talks at the Bomas of Kenya, a tourist village, and unveil a draft constitution which bore little relation to what had originally been proposed. The setting aside of ethnic rivalries, hailed as marking the Kenyan political class's coming of age, had outlived the elections only by a paltry couple of months. No sooner had the Mount Kenya Mafia climbed the ladder than they were kicking frantically away at it to ensure no one came up behind.

In State House, the process of ethnic polarisation was palpable. Since starting his new job, John had made a conscious effort during working hours to use Kiswahili – the national language – not Gikuyu, as would feel natural with tribal kinsmen. He knew how easily non-Kikuyu colleagues could be made to feel boxed out. The Mount Kenya Mafia showed no such restraint, finding his self-discipline quaintly amusing. 'We know you have a problem with this, John,' they would laugh, lapsing into a throaty barrage of Gikuyu. John would shake his head at the message conveyed. 'I used to warn them: "This talk will fix us."' He noticed how mono-ethnic State House had become. 'When meetings took place, they would all be people from the same area. All the key jobs were held by home boys.' The old tribal rivalry had returned – or rather, John realised, it had never actually gone away. 'With the collapse of Bomas I realised we had never been serious about power sharing. Kiraitu Murungi, the very man who had written about the problem of ethnicity, was the first to use the term "these Jaluos" in my presence.'

At a formal dinner in London several years later, I found myself discussing with John and a British peer of the realm, in light-hearted vein, what were the little signs that betrayed the fact that once-reformist African governments had lost their way. 'My measure is the time a person who's agreed to an appointment keeps you waiting,' said the Lord. 'If it's half an hour or under, things are still on track; more than half an hour and the place is in trouble.'

I quoted a journalist friend who maintained that the give-away was the moment a leader added an extra segment to his name – 'Yoweri *Kaguta* Museveni', 'Daniel *Toroitich* arap Moi' – but added that I regarded the size of the presidential motorcade as the tell-tale indication that the rot had set in.

John had been silent till then. Now he suddenly spoke up. 'How about the time it takes for the man in charge to get a gold Rolex?'

'But surely Kibaki already had a gold Rolex?' I asked, surprised.

'Yes, but this was a brand-new one. Very slim, with a black face and diamonds round the edge. It was so new it hadn't yet been measured to size, and it dangled off his wrist. That's why I noticed it, because it didn't fit.'

'So, then, how long did it take?'

'Just three months,' John said, with a grim shake of the head. 'Just three months.'

6

Pulling the Serpent's Tail

'KANU handed us a skunk and we took it home as a pet.'

JOHN GITHONGO[17]

In April 2004, Kenyan MP Maoka Maore received a mysterious phone call telling him that if he visited a fellow MP from the tea-producing area of Limuru, he might find some interesting paperwork there.

Maore would subsequently discover that at least five other MPs were already in possession of the same documents, which someone – almost certainly a disgruntled corporate executive – was energetically leaking. Fearful of the implications, none had acted. But Maore, a cheerful scallywag with a taste for the limelight, was made of more daring stuff. Proud of the role he had played in a 1994 exposé of kickbacks paid during the construction of an airport in Moi's home town, he boasted that his name struck fear in government circles. Expose one scandal, he had discovered, and all sorts of people will approach you with incendiary information about others.

The tip-off whetted his appetite. Rumours had been swirling around the Kibaki government for months, involving the procurement of AK47s, handcuffs and police cars. An administration which had vowed to crack down on graft had itself, it was said, begun 'eating'. Once he got his hands on the papers, he immediately tabled them in parliament, not entirely certain himself what they revealed.

The first document was a copy of a 2002 tender opened up by the previous government to supply Kenya with a computerised passport printing and lamination system. Nothing strange there – in the wake of Al Qaeda's 1998 bombing of the US embassy in Nairobi, Washington had been pressing Kenya, seen as a soft target for Islamic extremists filtering in from Somalia, to upgrade its passport system and better monitor its borders. The highest bid for that tender had been made by De La Rue, a British company, while the lowest came from Face Technologies, an American firm. What was strange, if the second document Maore tabled was to be believed, was that the tender had gone to neither. A payment voucher showed a Central Bank down-payment to a British rival called Anglo Leasing and Finance Company Limited.

This contract, which had never been put out to competitive tender, was a bloated, murky thing. For one thing, it was worth $34 million, nearly three and a half times as much as the lowest bid made back in 2002, which the government would ordinarily be expected to accept. What was more, the company awarded the contract, Maore reminded colleagues in the House, hardly boasted a savoury reputation. Six years earlier, under the former KANU government, Anglo Leasing had been blacklisted for supplying Kenya's police force with over-priced Mahindra jeeps – 'a cross-breed between a tortoise and a snail', in the words of a local newspaper – which broke down so regularly the police became a laughing stock. It looked as if officials at the ministry of home affairs had approved a contract inflated to the tune of at least $20 million. The whole deal gave off a sour, suspicious smell.

As far as the public was concerned, Maore's parliamentary question marked the start of the Anglo Leasing affair, the Kenyan equivalent of the break-in at the Democratic National Committee headquarters in Washington's Watergate complex. Today Maore marvels at what followed from his moment of chutzpah. 'It was like a dream in which you pull the tail of a snake, you keep pulling, and you find that it just goes on and on forever.' For John Githongo, however, Maore's action brought into the open an issue he had been probing for six long, anxious weeks, but naïvely believed he had brought under control. 'I thought I had it contained. We'd been trying to quietly fix the problem behind the scenes. Then, suddenly, the cat was out of the bag.' He would later come to feel a certain gratitude to Maore for exposing a matter which would prove too big for a mere permanent secretary. But at the time, convinced this was a minor affair that could be dealt with discreetly, the MP's intervention was just another problem to add to his growing number of headaches.

John, too, had been hearing rumours of new graft, of dodgy procurement contracts and lavish spending by members of the NARC administration, who had been buying up large villas in Nairobi's most attractive suburbs. His colleagues, he registered with growing alarm, were changing as the temptations of high office came their way. Many had spent the 1990s in the badly-paid world of political activism, setting up NGOs, braving the GSU batons, enduring police harassment. While they had pursued the cause of multi-party democracy, they had watched less idealistic friends, focused on businesses and careers, overtake them, moving from scruffy areas like South C to the pristine gated communities of Runda and Muthaiga. Now came the chance to narrow that gap after the years of self-denial. 'I had friends who bought three separate properties at once. They were handing their wives $100,000 in spending money,' remembers John.

At TI-Kenya, Mwalimu Mati also noticed the flowering arrogance of an administration that had started out eager to collaborate with former colleagues in the human rights world. With the launch of various inquiries into graft out of the way, NARC saw itself as beyond consultation. 'At the end of the various commissions and task forces, civil society stopped being involved. The reports were being given to the minister and president and dying a death. In the first six months to one year, people started making excuses. And then it was: "Butt out, *we're* the government."' The same men competed with one another to see who could secure the biggest office, the most ostentatious car. The Kiswahili term for the moneyed elite is '*wabenzi*' – a reference to the Mercedes Benz beloved of VIPs the world over – and

NARC officials wasted no time in confirming its literal accuracy. In their first twenty months in office, government officials spent at least $12 million (878 million shillings) on luxury cars, a survey by the Kenya National Commission on Human Rights (KNCHR) and Transparency International revealed. The sums spent on E-class Mercedes Benzes, top-of-the-range Land Cruisers, Mitsubishi Pajeros and Range Rovers could have provided 147,000 HIV-positive Kenyans with anti-retroviral treatment for a year. 'There was something of the Scarlett O'Haras to the Kibaki government at that time,' chuckles a Kenyan Asian businessman friend. 'They were gathering their flouncy petticoats around them and declaring: "As God is our witness, we'll *never* go hungry again!"'

The realisation of the virtual impunity enjoyed by those with connections to State House was hitting home. It had the giddy impact of a sudden rush of blood to the head, the first sniff of cocaine to the novice drug-taker. No longer ordinary mortals, they had become supermen, invincible, omnipotent. 'It's completely intoxicating, mesmerising. I could see it in their eyes,' remembers John. 'It's a point you reach. You simply do it because you can.'

Kenyan wags, the anti-corruption chief knew, had begun joking that the acronym NARC stood not for 'National Rainbow Coalition', but for 'Nothing Actually Really Changes'. Political commentators were reporting that the president's coterie had capitalised on his stroke and consequent inattention to get up to all sorts of mischief. Alarmed by a tangible sense of drift in State House, John confided to his diary that it might be time to consider resignation. But he stayed his hand. If he hadn't believed it was possible to reform a system from within, after all, he would never have accepted the job in the first place.

Trying to probe the provenance of all this easy cash, John found there was a striking difference in the treatment he now received from Kenya's National Security Intelligence Service (NSIS), which had been so very helpful when it came to dusting off the skeletons of the former regime. While the service had fallen over itself to provide information on Moi-era sleaze in the early days of the NARC administration, it proved a different matter when it came to the new government's actions. While superficially friendly, meetings with the Kenyan intelligence services were little more than exercises in futility. 'Their reports were complete rubbish, totally useless and unhelpful.' If John was going to do a decent job of policing his own government and not just pursuing the outgoing administration, he gradually realised that he would need to find his own, independent sources.

He was not on totally unfamiliar territory. During his time at TI, John had occasionally paid for information when compiling reports, so he had had some experience in recruiting sources. His natural propensity for befriending everyone and anyone, his ability to make both office cleaner and VIP feel equally appreciated was, as it happens, the spy recruiter's most treasured skill. Setting up a mini-intelligence network to rival the NSIS was never his intention; a policy of desperation, the thing began almost of its own accord.

From the start he'd operated an open-door policy, making clear to all that anyone – whether civil servant, politician, military officer or private businessman – was free to walk into his office at State House with useful information or to voice concerns. 'I didn't need to go looking, people would come to me.' To those who took up that open invitation, appearing at his doorstep with relayed rumours or suspicious documents that had passed across their desks, John was gently encouraging, gradually building up a relationship of trust. 'I'd say: "Gosh, you have this. That's really very interesting. But I think there's a letter missing here . . ." And they would go off to find the letter.' He focused on the departments which held the most power, where the most egregious offences seemed likely to occur: the Office of the President, the finance ministry, the ministry of internal security. Having sowed the seed, he waited to reap his slow harvest. 'The trick, I found, was never to be in a hurry, never get excited.'

What makes a law-abiding functionary, hardly the devil-may-care type, lift his or her nose above the daily grind and turn sneak, risking exposure, prosecution and dismissal? It was never for the money, something which was only mentioned late in the process, and usually at John's insistence. For many of those who would become his *de facto* informers, a profound and justified sense of betrayal explained the readiness to help. They were Kenyan voters too, after all, and like the mass of the populace, had believed NARC when it had promised a new dawn. Kibaki's campaign rhetoric had been almost too effective – they had taken the anti-graft message to heart. Now they knew, with a certainty not available to the ordinary Kenyan, that the old games were starting up again. Nothing smarts quite like the dashing of raised hopes. It forces the deluded to regret the best part of their nature: their readiness to believe in a better world. They felt they had been made fools of, and they wanted other Kenyans to know what was going on. 'Some were very angry,' remembers John. 'They'd say to me: "This used to happen under Moi. If you let this get out of hand, we've seen what happens. We're glad you're here."' And it was easy to rationalise the move from dutifully cooperating with the 'Anti-Corruption Czar' as he was now known, just as they had initially been explicitly instructed to do by the president himself, to slipping that same individual – such a likeable young man – information they knew in their hearts their superiors wanted kept secret. Only a tiny step.

John reserved his keenest attention for the disappointed and downtrodden, for those whose careers were going nowhere. 'I'd look for people who were frustrated. The employee who hadn't been promoted for a long time, the one who never got sent on training workshops, who couldn't get *per diems*, was being sexually harassed by the boss or felt that promotion had been denied on purely ethnic grounds.' For the man who had charmed Nairobi's donor community, winning the confidence of a grim-faced factotum in a worn nylon suit and broken-down shoes, the pre-retirement functionary who lunched off a cup of sweetened tea, took shelter under a plastic bag when the rain pelted down and went home to meals bulked out with the ubiquitous *sukuma wiki* – Kenyan greens – came all too easy. Bullied bearers of messages, loyal takers of notes, conscientious file-keepers, these ignored drudges bore astute silent witness to every top-level sleight of hand. It was the technique used by the CIA when it infiltrated Congo's political circles in the 1960s by paying waiters to relay bar talk, and Rhodesia's guerrilla movements when they used herdsboys to report on army movements. The perfect informer is the employee so insignificant he or she has become part of the furniture, literally invisible to those in power.

And there was another motivation for these moles in the making: the sheer excitement John represented in otherwise humdrum lives. 'To the uninitiated, the secret world is of itself attractive,' John le Carré wrote in *The Little Drummer Girl*. 'Simply by turning on its axis, it can draw the weakly anchored to its centre.' For those whose cramped lives were hemmed in by their rank, their insignificance spelt out in gradings, job titles and pay structures, there was an extraordinary thrill to be had in the knowledge of being inside the loop, invisibly pulling the strings of those who made their working lives such a chore. 'I may look like nobody, yet I know more than you,' they whispered to themselves, hugging that secret knowledge – simultaneously revenge and compensation prize – to their chests as the loud men in suits, gold Rolexes peeping below their cuffs, threw their weight around. 'If I want, I can fuck with you.'

As the communications got more sensitive, it became dangerous to be seen in John's company. Many of his sources confined themselves to text messages, or calls to his many mobiles. He owned over a dozen, their chargers overwhelming every power point in his Lavington home. He worked them as an experienced conductor leads an orchestra, each player given his or her proper place in the musical score. Numbers were distributed in categories, so that he knew who was likely to be calling before he lifted the handset. He came to know which informants would ring at which time – if it was 6 o'clock in the morning then it must be so-and-so, calling before going to work – who would make contact once a week, who wanted a daily chat and who would only surface sporadically with something really important.

Picking up the documents was the tricky part. They met in bars at 2 o'clock in the morning, in the forecourts of deserted petrol stations, in private homes – always at night. When John resorted to couriers, they were individuals he knew and trusted, and instructions were kept to a minimum. 'If you go to this shop, you will find a certain document waiting for you there.' If the matter was really sensitive, he would wait for the civil servant to take annual leave, drive to his upcountry *shamba*, and pick up the document in person.

Only John knew the names of his informants, which were never written down. To protect them, their payments – recorded in a special book stored in his State House office safe – were listed under code names: A2, B1, D3 . . . By the end, John was running a small stable of around twenty informants, their fees ranging from 50,000 to a hefty 200,000 shillings ($2,560) a month. It was a job that played to his talents, this man with a natural bent towards intrigue and an appetite for gossip. But it was not one he enjoyed; it made him feel shifty, tainted. 'I was spying on my own colleagues, which didn't feel like an honourable thing to do.' He was angrily aware that had Kenyan intelligence been delivering, none of it would have been necessary. But as the reports came in, it was ever clearer that he had done the right thing in striking out alone. 'Someone would tell me something. I'd ask the Kenya Anti-Corruption Commission to investigate and the same thing would come up. Then the attorney general would investigate, and the same thing would come up again. So I would ask myself: how come only Kenya's four-billion-shilling-a-year intelligence service can't get this right? The intelligence service was part of the architecture of corruption, and I had no confidence in it.'

Since early March 2004, John's informants had been talking to him about the Anglo Leasing contract. Two businessmen with spotty reputations – Deepak Kamani, whose Kenyan Asian family had been linked with the Mahindra jeep débâcle, and Jimmy Wanjigi, son of a former cabinet minister – were being cited in connection with the deal. But the other names mentioned were worryingly highly placed. Top of the list came vice president Moody Awori, minister for home affairs, and Chris Murungaru, minister of internal security in the Office of the President. Two permanent secretaries' names were cited – Dave Mwangi at the ministry for internal security, and Joseph Magari at the ministry of finance – as was that of another top civil servant: Alfred Getonga, Kibaki's personal assistant. Many of those named were the very men who had stepped forward after the president's stroke, effectively running the country during his illness. The roll-call went to the core of the Mount Kenya Mafia.

What was worse, John's sources were telling him something that Maoka Maore did not know. Anglo Leasing and Finance was no more than an address in Upper Parliament Street, Liverpool, an empty title; and since there was nothing behind that name, it seemed reasonable to assume that Kenya would never receive the tamper-proof, fraud-resistant passport printing systems Anglo Leasing had just been contracted, at inflated expense, to supply.

Before Maore addressed parliament, John had tried raising the matter with Moody Awori, in whose docket immigration, and anything to do with passports, fell. 'Uncle Moody', a politician from western Kenya with a liking for flowing robes and wide hats, enjoyed the image of a kindly elder statesman. But he proved distinctly evasive, telling John he made a point of not knowing anything about such things. 'Sticking his head in the sand,' John noted in his diary.

Anglo Leasing was not the only procurement deal John's informants had mentioned as suspicious. Another contract involved the construction in Spain of a frigate for the Kenyan Navy. Several companies had provided quotes, but the contract went to another outsider: a Cypriot of Sri Lankan birth called Anura Perera who had done a great deal of business with the Moi regime. Perera's tender had been far higher than any of the other companies', but he had nonetheless won the bid. The issue did not make a big impression on John until a week later, when the finance minister David Mwiraria pulled him aside at a conference to warn him that Murungaru was asking whether John had authorised an investigation into Perera's bank account. 'Mwiraria whispered to me that Perera was a strong supporter of the president and had backed him for over ten years and had even paid the president's medical bills incurred in London following his road accident in 2002,' he noted.

This was not exactly encouraging news for the man whose job was to liaise with the president in the fight against sleaze. It was with a distinctly uneasy mind that John briefed the president the day after Maore's parliamentary performance. Anglo Leasing, he warned his boss, was likely to prove the first big case of graft within the NARC administration. Given the level of unhappiness created by the botched constitutional review process, which left NARC vulnerable to charges of ethnic favouritism, it was vital the government demonstrate that it was devoting as much energy to pursuing graft within its own ranks as it had to exposing sleaze under Moi. Kibaki seemed unsurprised by both the scam and the names being mentioned, but agreed the Kenya Anti-Corruption Authority (KACC) should be activated. The structures to which NARC had pinned its anti-sleaze credentials were about to be put to the test.

The president had little choice, for parliament was now on the case, KANU MPs, and even some of their notional rivals, licking their lips at the scent of NARC blood. The head of the National Security Committee, NARC's David Mwenje, arrived in John's office and announced that he intended to summon various politicians and officials to give evidence. He was carrying a stash of documents, whose contents – relayed by John's spy network – were already familiar to the anti-corruption chief. Someone had been leaking with abandon.

Having set the ball rolling, John left with Mwiraria and Kiraitu for London, where he had an appointment at the offices of Kroll, the risk-consultancy group investigating Goldenberg on the Kenyan government's behalf. While the three Kenyans were there, John took the opportunity to ask for British records to be checked for evidence of Anglo Leasing. The checks, conducted on the spot, confirmed what John's sources had said: Anglo Leasing was not a legal entity. Aside from an address in Liverpool, there was, in fact, a baffling absence of information about the mysterious company which had had so many dealings with the Kenyan government. The blatancy of the scam seemed to embolden justice minister Kiraitu, who was in take-no-prisoners mode on the plane returning to Nairobi. 'The time has come for big heads to roll,' he declared. John was delighted. 'If we act firmly on this, the rest of our term will be a free ride politically. We will occupy the moral high ground,' he told his colleague. Stamping out graft was not only the right thing to do, it made political and electoral sense.

The return to Kenya was like a cold shower. Fired up, John couldn't wait to clear airport immigration to get things moving. He rang the head of the KACC from the VIP lounge at Jomo Kenyatta. The news was grim: the head of the civil service, Francis Muthaura, was refusing to cooperate with investigators until parliament had completed its own inquiry. Buoyed by his defiant stance, the permanent secretaries at Finance and Home Affairs were also stonewalling, with the former refusing to give the KACC a copy of the original Anglo Leasing contract. Chris Murungaru, minister for internal security, was also cutting up rough.

But a 2 May exchange with the president reinvigorated John. Kibaki, in fighting spirit, said he knew that Alfred Getonga, Murungaru and Jimmy Wanjigi were all involved in Anglo Leasing. The money was probably already 'eaten', he acknowledged, but John must press ahead as fast as he could. 'Proceed without mercy,' Kibaki ordered. His anger at the gathering revelations certainly seemed to confirm the scenario of a high-minded leader betrayed by shifty aides, a scenario John was more than happy to embrace. With Kibaki apparently backing him to the hilt, John redoubled his efforts. All further payments to Anglo Leasing, he wrote to Magari, must be stopped.

John had not expected his zeal to endear him to his ministerial colleagues, given the extent of the network his informers had sketched. It was just as well. On 4 May, vice president Moody Awori invited him for a lunch of stew and chapattis at the vice president's villa in Lavington, where they were joined by Kiraitu and Murungaru. Once the guests had arrived, 'Uncle Moody's' genial expression suddenly vanished and he turned on John. 'Now, what's all this about?' There was no need to investigate Anglo Leasing further, he insisted – he had already explained the matter in a statement to parliament. When John disagreed, a tense discussion ensued. Recording the encounter in his diary, John had a surreal sense of priorities being inverted. Anglo Leasing, for these men, was not the issue. *He* was the 'problem' they had all gathered to resolve.

Perhaps the element of the lunch that pained him most was the new stance adopted by Kiraitu. John knew enough about the Harvard-educated minister's background to hold him in considerable respect. Kiraitu had been one of a group of pro-opposition lawyers who had braved the wrath of the Moi regime in the 1980s, fighting for multi-partyism, defending political detainees. A member of what the media had dubbed the Young Turks, he had been harassed, monitored and followed before finally going into exile in the United States. There was something about Kiraitu's face, with its soft and fleshy lips – lips that would have been alluring on a woman but looked slightly repellent on a man – that suggested weakness, and in many of his public statements he had revealed a startling crudeness. His long friendship with Chris Murungaru, whose name surfaced with monotonous regularity every time Anglo Leasing was mentioned, was also a source of concern. But Kiraitu still stood for much that John believed. What had happened to the enthusiasm he had shown on the flight home from London?

It was about this time that John systematically stepped up a practice he had initiated the year before. This activity would later trigger his critics' most vitriolic abuse and leave even otherwise sympathetic Kenyans shaking their heads: he began secretly taping conversations with his colleagues. Why did he enter into what he would subsequently acknowledge was 'morally disastrous territory, the worst form of betrayal, the most discomfiting thing I've done in my entire life'? Initially – how ironic this would come to seem – it was simply in order to be able to prove his *bona fides* to the boss. In the first year of his tenure, the most contentious material he heard came from businessmen passing through his office. Without any paperwork to prove his claims, he realised that if he relayed these gobbets of information to the president, only for those who had provided them to think better of their frankness, he could be made to look either a villain or a fool. The systematic taping would prevent him from being stitched up in front of the president. Things said in John's presence could not later be denied. He had told the *Mzee* what he was doing from the start, leaving it to Kibaki to decide whether to pass that information on to other State House players. The president had reacted in his usual laid-back fashion. 'He just laughed.'

But as the Anglo Leasing scandal began to unfold, the motives behind John's taping subtly changed. It was no longer merely a question of convincing his employer, but of justifying himself before history. As the disconcerting admissions piled up and his suspicions about his closest colleagues mounted, the cold realisation of just how much he stood to lose – reputation, credibility, employability – dawned. Listening to the Goldenberg hearings he had helped engineer, John had imbibed a pertinent lesson: throughout Kenyan history, civil servants had always served as scapegoats, fall guys, when high-profile financial scandals came to light. The elected politicians who issued the orders had always walked free, the size of their ethnic constituencies, in a world of political alliances, coating them in Teflon. To his own ears, the hints that were now being dropped were so shocking in their implications no sane person, he felt sure, would believe him if they did not hear the words for themselves. He would be mocked as a paranoid fantasist, a Walter Mitty character whose mental instability had, sadly, not been spotted by his recruiters. The recordings – his 'wires', as he called them – were all that stood between him and ignominy. 'By 2004 I knew the wires would be the only thing to save me.'

If he found what he was doing repugnant, there was a sense in which he was using what small and idiosyncratic weapons he had to combat vastly superior forces. The Mount Kenya Mafia might have a huge network of civil servants, intelligence agents, generals and police chiefs to do their bidding, but there were areas in which they were surprisingly weak. The kind of man who gets nerdish pleasure from keeping up with the latest computer software, music gadgetry and mobile phone special features, John had no trouble working out how to download digitalised sound, set a tape recorder onto time-delay voice-activation or encrypt his internet traffic to shield it from prying eyes. Many of those he was dealing with belonged to a generation of technology-allergic old-schoolers who prided themselves on barely being able to type – that was a secretary's job – hardly grasped the concept of voicemail and stared at their mobile phones in bemusement when they bleeped. 'I had an advantage. I'm a technology geek and these were guys who had trouble sending an SMS.'

There was another side of John to which the taping appealed. Since his schooldays, he had been recording his own life with a scientific thoroughness more typical of a zoologist than a diarist. He had taken to heart Socrates' maxim that the unexamined life is not worth living. The wires were just another ingredient in the process of existential reckoning being conducted by this accountant's son. Who, in his mind, would eventually tot up the pluses and minuses and come up with a definitive balance at the end of his turbulent life story? God? John himself? The Kenyan public? Future historians? There was no clear answer, but that didn't make him feel any less compelled to log and note. Capturing events on paper and on tape was the one way he could impose order on the chaos of events swirling around him. Just as it was the old timers' misfortune to be dealing with a technogeek, it was also their bad luck – in a political world which relied for its protection on word-of-mouth instructions, whispered consultations and the absence of a paper trail – to have recruited a man who felt compelled to record every daily incident for posterity.

What happened next revealed the sensitivity of the Anglo Leasing affair. In the early hours of the morning, the phone would suddenly ring at John's home. When he lifted the receiver, an anonymous voice, growling in Kiswahili, would warn him his life was in danger, promise he would soon see who he was dealing with, and that he would regret what he had done, bastard, cheat and liar that he was. At other times there would be nothing but a brooding silence on the end of the line. And when John hung up, the same thing would happen again – call, answer, silence, call, answer, silence. Their very relentlessness provided a clue as to who lay behind the calls. It bore all the hallmarks of a scare-off operation by Kenyan intelligence.

He did not mention these calls to Kibaki. It came with the territory, he reasoned – no point going whining to the boss. But within days, a campaign of systematic disinformation was being piled onto the death threats. The aim was to sully John's reputation to the point where his professional actions would have no public credibility. The shadowy figures who had set themselves the task of smearing him chose a classic route. John might have a girlfriend of four years' standing, the loyal Mary Muthumbi, but he was that rarest of creatures: a thirty-nine-year-old African bachelor. Why exactly *was* he still unmarried and childless at what, by Kenya's standards, counted as a strangely advanced age? With a bit of imagination, those simple omissions could be transformed into out-and-out perversion.

The trick of the successful smear campaign is to include just enough fact – real nails on which to hang a work of bold invention – to persuade an audience that 'there must be something in it'. St Mary's, John's old school, had, in its small way, something of a gay reputation. As Alfred Getonga and Jimmy Wanjigi could not help but be aware – they were alumni, after all – rumours had swirled around the Catholic priests teaching there. *Voilà!* The weapon presented itself. Editing out Mary's role in John's life – everyone knows, after all, that a girlfriend or wife's existence proves nothing when it comes to these *perverts* – Nairobi's gutter press unleashed a barrage of scurrilous stories. One of John Githongo's lovers, it was said, was a male journalist at the *Nation*. The couple lived together as man and wife. Another supposed boyfriend, a white rally driver, had committed suicide when John broke off the relationship, driving out to Nairobi's National Park, hooking a hosepipe up to his car's exhaust and gassing himself. The underpaid reporters on Nairobi's gossip sheets, always vulnerable to the proffered brown envelope full of cash, wrote breathlessly of clandestine homosexual trysts, of hired prostitutes – this time heterosexual – and hinted at mini-orgies organised by a hitherto-unknown group called the Royal Gay Society, the baroque 'royal' ingredient no doubt intended to whip up anti-imperial sentiment. An anonymous leaflet circulated parliament, making similar claims.

Only partially aware of what was going on, members of John's own team reacted to the slurs with hilarity. 'Oh, the gay thing was just hysterical,' guffaws Lisa Karanja. 'After he showed me the cuttings, I decided to make him a crest, of a family shield with two pink flamingos and the initials RGS – for Royal Gay Society – on it.'

The claims barely raised an eyebrow among the foreign diplomats, journalists and aid officials who worked in Kenya. This community of Western sophisticates wouldn't have minded if John had been a feather-boa-wearing transvestite in his free time, so long as he wasn't on the take during professional hours. But John's enemies knew their audience. In much of Africa, 'homophobia' is a meaningless term, given the depth of public hostility to homosexuality. The practice is illegal in Kenya, and for a vast swathe of the prudish public it represents the ultimate of depravities, a vice imported from the effete West to corrupt manly African youth. Kenyans will ask you, pop-eyed with disgust, in the incredulous tone of voice they would adopt towards a report claiming goats had the vote in California: 'Is it true that in your culture homosexual people can *marry* one other?' The gay barb hit home, leaving an indelible stain, just as its authors had known it would.

Aware that friends and family were reading the smears, John cringed, but there was nothing he could do. As a permanent secretary, he'd automatically been assigned a bodyguard, but it had been a relaxed arrangement, with his minder clocking off at the weekend. The man who looked so much like a bouncer himself now asked State House for extra protection, and was assigned two bodyguards who accompanied him everywhere. So free hitherto, he now had constant chaperons.

A whispering campaign had started in State House, with John – known to be on good terms with journalists – widely blamed for leaked stories appearing in the Kenyan press. His colleagues' hostility had become evident. But perhaps it was worth it, for his efforts seemed to be paying off. On 6 May, Chris Murungaru announced in parliament that the government had cancelled the Anglo Leasing contract, pending inquiries. A week later, when the KACC and the controller and auditor general confirmed in separate reports that Kenya had fallen victim to a gigantic attempted fraud, president

Kibaki agreed without hesitation to John's recommendation – as he handed over his final report on the affair – that the various suspect officials be sacked. 'OK, that's it!' he told him. This, John felt with a glow of pride and relief, was what it was to be backed by your head of state: all became straightforward and clear. As John took his leave, the president asked him to summon Francis Muthaura, head of the civil service, to State House – presumably in order to inform him of the looming staff changes.

And then things got strange. Strange enough to make one wonder what actually took place between the president and his head of civil service once John left the room. Later that same evening, a jubilant Muthaura rang John to tell him that Anglo Leasing had been in touch, promising to refund the government's 91-million-shilling downpayment. He announced this with the air of someone who had consigned a troublesome problem to history. How a non-existent company could suddenly find itself a voice, and why that phantom entity should then choose to call the head of the civil service rather than the contract's signatories, remained unclear. Sure enough, the Central Bank confirmed the refund had gone through. A few days later, on 17 May, Joseph Magari and Sylvester Mwaliko, permanent secretaries at the finance and home affairs ministries respectively, were suspended, furiously protesting their innocence. Along with them went Wilson Sitonik, head of the department vetting all computer procurement for the government.

It was progress of a sort, but John quickly realised that while he considered these preliminary steps in the right direction, those around him regarded them as sops with which to silence an increasingly irritating, overzealous colleague. The money had been returned, end of story. Kiraitu, who had once been so keen to see heads roll, wandered into John's office unannounced, looking distracted, and expressed the hope that the investigations would end with the Anglo Leasing refund. People, said the justice minister, were beginning to wonder if John appreciated the political costs of his work. That same day, a palpably nervous Mwiraria popped by to warn John that Jimmy Wanjigi had sworn to kill him – which raised the obvious question of how the finance minister knew this. But John could not ease up. He was now receiving reports of a *second* suspect Anglo Leasing deal, this time for the notional provision of a state-of-the-art police forensic laboratory. His informants were telling him of other dodgy security contracts, too, to which Anglo Leasing's name was not attached. Like Hercules, he had cut off one of the hydra's heads, only to find a host sprouting in its stead.

To those watching John, it must have been clear that the death threats and smears had failed: his determination to pursue Anglo Leasing had not dimmed. A new approach was needed. On 20 May 2004 he was summoned to the justice minister's office. 'The general message,' he told John, was: 'Tell Githongo to go a bit slow.' When John's response was uncompromising, Kiraitu revealed his hand. Opening a drawer, he pulled out a file he said had been given to him by a Mr A.H. Malik. That name was familiar – Malik was a Nairobi lawyer who had loaned John's father money in the 1990s, when Joe Githongo planned to develop a piece of land on the outskirts of the city. It was one of his father's worst business moves: he had been unable to repay the loan, Malik had gone to court, the court had found against Joe, and the dispute festered on. What had any of this to do with John? The money loaned to his father, Kiraitu claimed, had not come from Malik at all, but had originated with Anura Perera, the Cypriot businessman behind the suspect frigate deal, the kind friend who had paid Kibaki's London hospital bill. There seemed no getting away from this generous individual. So John's father was in Perera's debt, but Perera, Kiraitu said, would be happy to settle the dispute amicably so long as John agreed to turn off the heat.

John sat blinking, struggling to digest the implications of this conversation. The first thing that astonished him was the extent and organisation of the plotting against him. He could only guess what measures Perera had taken to win ownership of the Malik loan, for he didn't believe for an instant that his father had originally, without realising it, borrowed from Perera. The second was the clumping crudeness of the overture. The man entrusted with Kenyan justice, who he had regarded until now as an ally in his war on graft, had just tried to blackmail him. Having failed to sway John with threats, smears and appeals, the establishment had resorted to the most cowardly of routes – trying to get at him via his father. He struggled to conceal his anger and disgust. 'Let the real principal show himself,' he told Kiraitu curtly, 'and I'll deal with him directly.'

One of the problems John was experiencing was the rush of events, so fast and furious he barely had time to analyse them. He received a call from a director in Kenyan intelligence. The man sounded agitated, but there was something contrived in his manner. His colleagues, he said, had hatched a scheme to discredit John. Three prostitutes – two women and one man – had been paid to demonstrate outside police headquarters claiming to have had sex with John and been poorly treated. The director said he was 'working very hard' on John's behalf to contain the situation. Alarmed, John sent a staffer to meet the prostitutes and photocopy their IDs, so that he could find out more about these supposed witnesses. Then he paused, thought again, and let the matter drop. Treating the matter seriously, he realised, would be to play his harassers' psychological game. This was a distraction, meant to divert his attention from Anglo Leasing. He would not give his enemies that satisfaction.

John's first appearance before parliament's Public Accounts Committee to be questioned by MPs on Anglo Leasing almost felt like a relief. He was determined not to lie on the administration's behalf, and it gave him a chance to offload some of what he knew. But the relentless pressure was having an impact. He felt stressed, worn out. Even on nights when no one called to make anonymous death threats, he slept fitfully, waking repeatedly. He kept the details of what was going on to himself. Even if the Official Secrets Act hadn't barred government officials from discussing such matters with friends and family, he knew better than to expose his loved ones to the kind of threats he was experiencing. So toxic was the information he was gleaning, he felt he could not share it even with his staff. 'He wouldn't lie,' remembers Lisa Karanja. 'He'd tell you a lot was going on. But he wouldn't go into the detail.' She noticed his habit of repeating himself, as his mind, terrier-like, worked away obsessively at themes that niggled him. It was a sure sign of exhaustion.

When no one else shares one's perception of reality, only the madman believes he is *compos mentis.* From his ministerial colleagues to the head of the civil service, virtually everyone was telling John his zeal in pursuing Anglo Leasing was misplaced. For a moment, he wavered. Who was he to insist that it was, on the contrary, the only possible course of action? In something approaching despair, he texted Kiraitu to tell him he was easing off, as requested. 'That's good to hear,' Kiraitu texted back. 'Now life can continue.'

Studying that message, something in John jibbed. No, he would not oblige them all by going quietly. He would press on, he would continue the secret taping, but he must become more cunning. Let the Mount Kenya Mafia – he was confident Kiraitu would pass his message on to the rest of the group – believe he had surrendered. In public, he would adopt a softly-softly approach, winning some respite from the incessant sniping. The Big Man was not finished yet. In private, via his informant network, he would pursue Anglo Leasing with as much vigour as the need for secrecy allowed. And he would continue his clandestine taping.

On 2 June 2004, long after everyone else had gone home, Mwiraria, Kiraitu and John gathered in the finance minister's office to survey recent events. After the weeks of tension, mutual irritation and paralysing suspicion, it felt as though a tightened spring had suddenly been released. This was the calm that comes with trust, and the joint understanding that a serious crisis has been narrowly averted by dint of pulling together. At precisely this moment of assumed complicity, John's hidden tape recorder chose to start relaying its contents to the world at large.

He scrambled for the door, and returned expecting the worst. But fate was kind. The atmosphere in Mwiraria's office seemed unchanged. Kiraitu, in particular, was in meditative, confessional mood, speaking more freely than he had since the scandal broke because, John sensed, he was convinced the anti-corruption czar had seen the error of his ways. 'He admitted that he had not realised how high up and just how intricately involved members of our own administration were,' John wrote later that night in his little black book. He had to call on all his skills as an actor to conceal his dismay at what the justice minister said next. It was confirmation of a truth John had really, in his heart, known all along but had not wanted to confront. 'Anglo Leasing,' Kiraitu ruefully acknowledged, 'is *us*.'

It was an astonishing admission to make before the man who had been given the remit of eliminating corruption. Kiraitu's confident assumption that John would nod quietly in agreement, rather than leap to his feet and start working the phones in sleaze-buster mode, might seem bizarre to the outsider. In fact, it was based on one all-important fact, a keystone on which, in the eyes of the ministers and their colleagues, a solid edifice of cooperation and mutual protection could be built: John was one of them, John belonged. John was a Kikuyu.

7

The Call of the Tribe

'You're my older brother and I love you. But don't ever take sides against the family again.'

MICHAEL CORLEONE, in *The Godfather*

If you drive north-east from Nairobi, aiming for Mount Kenya, it takes a while to shrug off the city slums. Traffic slows to a crawl while doing its best not to stop entirely at Githurai roundabout, notorious for cut-throats and thieves, then bombs in relief down the Thika road. Some of the worst accidents in Kenyan history have occurred on this stretch of road, but few drivers let that deter them. Ordinary cars compete with crammed *matatus*, yellow jerrycans bobbing from roof racks like party balloons, to see who can flirt most outrageously with death while remaining on the road.

You know you've reached Thika when you start passing a series of wooden trestle tables, buzzing with tipsy wasps, where pineapples borrowed from the Del Monte plantation, a purple-grey expanse of scrubby sprouts, are cheekily sold. The motorway then crosses the swirling Chania river, whose falls are a favourite with lovelorn suicides, running briefly parallel to the railway. Then it is time to abandon the main road and turn west. If, seen from space, most of Kenya appears an arid expanse of yellow semi-desert, Central Province is its lush emerald kernel. In pre-colonial times, caravans heading towards the fearsome kingdom of Buganda would load up here with provisions. Awed by the farming skills of the locals, admiring travellers described these green valleys as 'one vast garden'. And that is still the impression today. The waters from two mountain ranges, Mount Kenya to the north-east and the Aberdares to the west, have carved the land into a series of moist valleys and mist-swathed ridges, the historic building blocks of Kikuyu society.

Take a detour to the top of one of these, and you will find yourself gazing across a vaporous Hobbitland of bottle-green dales, each ridge echoing the line traced by its neighbour until the blue layers, like a watercolour's delicate washes, blur into an inky distance. This is a man-friendly landscape of tamed shires, generous waterfalls and accessible horizons, a world away from the annihilating vistas of Turkana and Marsabit. Rich in laterite, the earth here is rust-coloured and sucking wet. After the rain, it boils thick and greasy under tyres and cars glide uncontrollably across its surface like drunken ice-skaters. When spattered on clothes it leaves indelible marks behind, like stale blood. So fertile is the soil, it's easy to succumb to a panicky claustrophobia when venturing onto the unmarked feeder roads, as the napier grass crowds in, cutting out the sunlight, making orientation impossible. 'True Kikuyu country,' gloats my Kikuyu driver as we slither past orchards of glossy coffee bushes and giant fronds of banana, fluttering like sails at sea. 'No one here will ever die of hunger, like in Ukambani.' But this giant allotment is straining under the press of population. It bears the marks of having been divided and subdivided, the strips of land that drape the hillsides like elastoplasts – each with its own *shamba* – signs of a paterfamilias's doomed attempt to do right by each member of an overly large family. The pressure on the soil here is so intense, the ability to coax life from the earth so instinctive, even road verges serve as vegetable plots, carefully tended seedlings growing within inches of speeding wheels.

Keep heading north on the main road, through meadows grazed by hobbled goats, and you eventually reach the market town of Muranga. It's easy to drive through this unremarkable place, perched on a rocky escarpment, without suspecting it was ever of strategic significance. But once, in an earlier manifestation as Fort Hall, one of the first British outposts, it played a vital part in a colonial empire's drive to pacify, occupy and settle East Africa's hinterland. A dozen kilometres beyond Muranga, a signposted dirt track veers off to the left and heads uphill, passing playgrounds of screaming schoolchildren in DayGlo jerseys, their bare feet coated in ochre. At the top of the ridge there is a sky-blue gate with the words '*Mukurwe Wa Nyagathanga*' – The Tree of Gathanga – painted upon it.

The compound inside may be officially gazetted as a tourist site, but it looks unkempt, virtually abandoned, the only sounds coming from the weaver birds chattering in the bush. The small office is locked up, leaving a posse of local villagers the task of hauling open the gates and showing visitors around. The mildewed skeleton of an unfinished hotel, intended to host the crowds some entrepreneur convinced himself would one day flock here, dwarves what those imagined tourists were meant to see: two traditional red-earth rondavels, huddled under a slim *mukurwe* tree which vaults across the clearing. The neglect is puzzling, given that this is supposed to be the spot where Kenya's biggest tribe first saw the light of day. This is the birthplace of a Chosen People, the Kikuyu nation's very own Garden of Eden, complete with symbolic Tree of Life. But then, the Kikuyu have always been a pragmatic people, their gaze firmly trained on the future, not the past.

According to the legend, an intriguing blend of history and religion, Kirinyaga – today's Mount Kenya, the peak the site looks towards – was the focal point of the Kikuyu world. Sitting just south of the equator, yet boasting a permanent giant moustache of snow, the summit was believed to be the seat of Ngai, God the Creator. The two licks of snow on the 5,200-metre peak, Kenya's highest, were said to be made from precious dust on which Ngai took his rest, and the mountain's name was derived from '*nyaga*' – 'ostrich' – a local bird to which the mountain, with its black volcanic base and white cap, bore a passing resemblance. Ngai created Kikuyu – the first man – in his own image, then took him up onto Kirinyaga to survey his future kingdom, a land whose forests teemed with fruit and rustled with wildlife, its valleys constantly watered by rivers from the mountain's permanent snows.

'Build your homestead on that spot where the fig trees grow,' Ngai told Kikuyu, pointing to Mukurwe Wa Nyagathanga. So Kikuyu settled here, and when he needed a helpmate, Ngai sent him beautiful Mumbi, the world's first woman. When the couple asked for children, Ngai sent them first daughters and then some handsome sons-in-law. It was from these youngsters' loins that the founding clans of the Kikuyu tribe sprang, one to each ridge. Each of the clans was named after a daughter – Wanjiku, Wangari, Wanjeri, etc. – for the Kikuyu were originally a matriarchal society. Quixotically, it was considered bad luck to specify how many daughters, or founding clans, the House of Mumbi contained. Whether discussing offspring, livestock or goods, it was safer to stick to a vague 'nine plus . . .', rather than a specific 'ten'. To quantify was to play into the hands of one's enemies, offering them potentially dangerous information.

Thus ran the myth, a monotheistic creation story with much in common with the one found in the Bible. Ethnologists tell a slightly different story. Along with a smattering of other Bantu tribes, the Kikuyu probably arrived in what is now Kenya after an infinitesimally slow migration that began in around 2000 BC in what are today's Nigeria and Cameroon. Responding either to the drying of the Sahara or the press of alien peoples, these Bantu communities arrived in Tanzania after tracing a continent-wide loop. Some then turned towards southern Africa, where their descendants were destined to become the Zulu and Shona. Others headed north-east, aiming for the coast before swerving back into the hinterland, where some settled in the Kamba hills. The group that became the Kikuyu called a halt in today's Central Province, edging out a local population of forest-dwelling pigmies. They had been wise in their choice of new home: these highlands were not only cool and fertile, they were located above the malaria line and were largely free of the tsetse fly, sparing them two of Africa's most devastating diseases. Numbers surged so dramatically that by the sixteenth century Muranga was unbearably crowded. Some then moved north to Nyeri, the well-watered area between Mount Kenya and the Aberdares. Others trekked south, crossing the Chania river and heading into the district of Kiambu, which now lies on the fringes of Nairobi, buying land as they went from the honey-gathering Dorobo tribe.

In his book *Facing Mount Kenya*, Jomo Kenyatta attempted to capture the essence of Kikuyu culture – fast becoming a romanticised memory – before it was lost to view, swamped by the white man's ways. Writing in 1938, when the Kikuyu population, today estimated at 7.4 million, was just one million strong, Kenyatta paints a rose-tinted picture of a stateless society in which extended families, known as *mbari*, lived in harmony on the ridges, herding goats, growing beans and brewing beer. Their tranquillity was disturbed only by occasional small wars with the nomadic Maasai, whose region stretched on either side of the Kikuyu escarpment. But the two communities' contrasting obsessions usually allowed them to rub along together peacefully enough. 'Wherever there is grass belongs to us,' was the motif of the cattle-herding Maasai, 'Wherever there is soil belongs to us,' said the agriculturalist Kikuyu.

This was a devout society, which respected the spirits of its ancestors while worshipping Ngai as supreme being. When the Kikuyu prayed they turned to face Kirinyaga, and sacrificed goats at the foot of sacred giant fig trees, nature's churches. It was a society which practised polygamy and marked the transition to adulthood with elaborate circumcision rituals which established special bonds of intimacy between members of the same age-set. The Kikuyu did not congregate in villages, and power was similarly decentralised, with councils of elders, known as *kiama*, taking key decisions and one generation passing responsibility to the next at a riverbank ceremony staged every thirty to forty years. Kikuyu warriors went about armed with spears and bows and arrows, but the community had no need of a standing army. The solitary individualism of Western thought could not have been further from the Kikuyu's collective vision of existence, in which a man's very identity was rooted in the group. 'Nobody is an isolated individual,' wrote Kenyatta. 'Or rather his uniqueness is a secondary fact about him: first and foremost he is several people's relative and several people's contemporary.'[18]

Crucially, Kenyatta also described a complex system of land ownership. Contrary to what the white settlers assumed, communal ownership of land was not a Kikuyu characteristic. Formally bought, carefully demarcated and privately owned, land was the bastion on which the tribal economy was founded. The ability to force the land to yield its riches was what made a Kikuyu superior, in his own eyes, to the feckless Maasai pastoralists who roamed the Rift Valley. 'There is a great desire in the heart of every Gikuyu man to own a piece of land on which he can build his home. A man or a woman who cannot say to his friends, come and eat, drink and enjoy the fruit of my labour, is not considered as a worthy member of the tribe,' wrote Kenyatta. Land not only conveyed status, it also provided a spiritual connection with past and future. 'It is the soil that feeds the child through lifetime; and again after death it is the soil that nurses the spirits of the dead for eternity. Thus the earth is the most sacred thing above all that dwell in or on it.'

It was this precious possession that colonialism placed in jeopardy, so perhaps it's no surprise that the Kikuyu showed themselves very far from docile in their early encounters with the white man. Count Samuel Teleki von Szek, a Hungarian explorer who led the first white expedition to northern Kenya in 1887, came under constant arrow attack while travelling through Kikuyu country, and claimed he had never come across more hostility during all his East African travels. A decade later, Francis Hall, the British District Commissioner after whom Fort Hall was named, found the Kikuyu 'exceedingly intractable' in the face of his attempts to hammer obedience home by setting fire to hundreds of their homesteads and confiscating tens of thousands of their goats and cattle. 'Too treacherous to be trusted to any extent, of a cunning, distrustful and treacherous nature, accustomed to look upon strangers as enemies,' he complained.

The implacable British Captain Richard Meinertzhagen, posted to Muranga in 1902, was also surprised by the ferocity of Kikuyu resistance. 'I must own, I never expected the Wakikuyu to fight like this,' he recorded in his diary after a successful punitive raid. He watched with sceptical disbelief as the first British settlers – including the ebullient Lord Delamere, most prominent of a clique of rollicking, black-sheep-of-the-family British aristocrats making Kenya their home – confidently started drawing up grandiose plans. These breezy new arrivals, regarded by a visiting young Winston Churchill as little more than 'ruffians', were determined to transform Kenya into 'White Man's Country', whatever the British government might feel about the matter. 'Sooner or later it must come to a clash between black and white. I cannot see millions of educated Africans – as there will be in a hundred years' time – submitted tamely to white domination,' wrote Meinertzhagen. And as he prepared to leave, he correctly guessed who would be the source of future trouble. 'I am sorry to leave the Kikuyu, for I like them. They are the most intelligent of the African tribes that I have met; therefore they will be the most progressive under European guidance and will be the most susceptible to subversive activities. They will be one of the first tribes to demand freedom from European influence.'[19]

With the arrival of the white settlers, life for the Kikuyu became increasingly bleak. The issue was not so much the *wazungu*'s confiscation of traditional Kikuyu land. According to the British Land Commission Report of 1933, whose charting of Kikuyu boundaries prior to white settlement has never been seriously questioned, only 6 per cent of that would ever be grabbed by the colonial powers, most of it in southern Kiambu. 'There is no doubt that the hardest-hit victims of land alienation were the Maasai and not the Kikuyu, the latter's clamour notwithstanding,' writes Kikuyu historian Godfrey Muriuki.[20] No, the *wazungu* represented a devastating challenge because they had effectively stolen the Kikuyu's future. Previously, the Kikuyu had always successfully negotiated access to an ever-widening area for their growing population with either the Dorobo or the Maasai. That territory had now been swallowed up by the White Highlands, future expansion permanently blocked. Forced for the first time to pay hut taxes by the 'traditional chiefs' imposed on them by the British, more and more young men migrated out of the

Kikuyu Reserve. They either sought work in Nairobi's expanding slums or became squatters, farming fields in the White Highlands that they could never hope to own.

Unlike the Maasai, whose rejection of modernity doomed them to marginalisation, many Kikuyu eagerly embraced the new ways, deciding that the route to success lay in adopting Christianity and Western customs. The *athomi*, 'people who read', replaced banana leaves on their roofs with corrugated iron, goatskins with shirts and trousers. Under the influence of the missionaries who had fanned across Kikuyuland they gradually abandoned polygamy and female circumcision, and insisted on learning English, language of the master race, rather than the Kiswahili the British thought appropriate. Writer Binyavanga Wainana pokes fun at these 'progressives', whose loyalty to the white man could be measured by the amount of Vaseline they used. 'You can see it in old photos: a generation of clean-cut, Vaseline Kenyans who had regular features, seemed to have no ethnicity, and carefully combed down their hair.'

Yet still they found the playing field pitched against them. In Kikuyu culture, the quality most admired is to be '*muthuri wirugamitie*' – an upstanding man, a man who earns his living from the sweat of his brow. Land ownership, traditionally, was what allowed a Kikuyu male to become captain of his destiny. Now many Kikuyu males found themselves demeaned. Unable to marry because they owned no property, thwarted in their desire to found family dynasties, they had assimilated faster than any other Kenyan community, yet what had this flexibility brought them?

Decades of grievances reached a head in the late 1940s, when a banned organisation, the Kikuyu Central Association, began secretly administering traditional oaths of loyalty to young Kikuyu, effectively signing up secret fighters for a coordinated campaign of civil disobedience.

As oathing quietly spread through the Kikuyu community, veteran activist Jomo Kenyatta returned from long exile in London to take the leadership of the Kenya African Union, an organisation pushing, through parliamentary channels, for black rule. When the British government refused in 1951 to bow to demands for the number of elected Africans on the colony's Legislative Council to be raised above five for a population of five million – as compared to fourteen for 30,000 white settlers – the possibility of compromise between settlers and Africans receded. One year later, alarmed by the growing number of attacks on white farms and the murders of suspected Kikuyu informers, Kenya's governor declared a state of emergency, deployed troops and arrested a hundred black leaders, including Kenyatta. It was a move which betrayed the degree of panic in the colonial administration. Despite time spent in Moscow, Kenyatta was no radical. He had so little sympathy for the revolutionary credo of the Land and Freedom Armies, the movement which would swiftly be dubbed 'Mau Mau', that its hardliners would discuss his assassination. The British decision to sentence this supposed ringleader to seven years' hard labour simply turned him into a national hero.

British press coverage of the Mau Mau rebellion would play on all the traditional Western stereotypes of the dark continent. This war in 'Terrorland', the British public was told, pitted plucky settlers' wives on lonely homesteads against a disturbingly irrational enemy in whose breast the Mau Mau's macabre nocturnal oathing ceremonies, involving animal sacrifice and, perhaps, bestiality, had awakened the most primeval impulses. The Kikuyu, it was said, had been plunged too suddenly into the modern world – 'Only fifty years down from the trees,' muttered the settlers[21] – and the jarring shock of the encounter between primitive culture and Western life had triggered some sort of psychological meltdown.

The details of Mau Mau's 'terrorist' atrocities were so gruesome – an elderly settler disembowelled in his bath, a tousle-haired six-year-old hacked to death amidst his teddies by the family servants – they obscured the reality of casualty numbers. The overwhelming majority would be black, not white. Historian David Anderson estimates that while sixty white civilians and the same number of British soldiers and policemen died during the insurgency, the number of Kikuyu dead probably reached 20,000.[22] For the Kikuyu community, a post-feudal society itself riven with inequalities and ripe for internal revolution, was tragically split. On the one side stood wealthy landowners who had prospered by collaborating with the British, mission-educated Christians who rejected Mau Mau's call for a return to traditional Kikuyu roots, and the Home Guard (subsequently renamed the Kikuyu Guard), a militia loyal to the colonial government. On the other stood Mau Mau's natural recruits: desperate young men, many of them landless squatters. Their oaths to Kirinyaga were marked by a cross of soil and animal blood smeared across their foreheads, and when fatally wounded in battle their last act, it was said, was to seize a handful of that same soil. This rich earth was what had nurtured them, the reason for laying down their lives, the element to which they returned in death.

As the campaign to suppress what was as much a civil war as an anti-colonial uprising gained momentum, with the Royal Air Force bombing and starving into submission two dreadlocked rebel armies mustered in the dank forests of Mount Kenya and the Aberdares, scarcely a Kikuyu family remained untouched. The Mau Mau did find recruits in other ethnic communities, with neighbouring Meru and Embu lending a particularly fervent hand. But Mau Mau was always predominantly a Kikuyu phenomenon, and that meant every 'Kuke' – the nickname alone sounded like a curse on settler lips – was suspect. In 1954, realising that Mau Mau cells in the capital were keeping the movement in the countryside supplied with weapons and information, the British rounded up some 15,000 Kikuyu in Nairobi. Those deemed suspicious were sent to bleak detention camps to be broken, 'cleansed' and rehabilitated, a process dubbed 'the Pipeline'. The rest were deported to newly-built villages in Central Province, complete with spiked moats and watchtowers, guarded by twitchy Home Guards. Families were often torn apart, with one son opting for life in the forest while his brother or father donned Home Guard uniform. So was the Kikuyu community as a whole, for it was noticeable that Kiambu, whose proximity to Nairobi meant its residents had been the first to be exposed to Western civilisation, produced precious few Mau Mau generals, while Muranga and Nyeri – more remote, less influenced by the white man –

produced the hardliners. The bitterness created by such divisions would rankle through the generations. The more grotesque the form violence takes, the deeper go its scars, and plenty of grotesque acts were performed during these dark days, on both sides. In the British detention camps, suspects were castrated, raped and beaten to death, while the Mau Mau decapitated, strangled and disembowelled suspected enemies and informers.

By 1960 the British authorities had won the battle, but lost the argument. A problem officials had confidently expected to last less than three months had dragged on for eight years. In that period, press coverage back home had changed beyond recognition, thanks in part to the efforts of Labour MP Barbara Castle and *Daily Mirror* journalist James Cameron. There was little sympathy for Kenya's settler administration in post-war Britain, where reports of Happy Valley debauchery had gone down particularly badly. Crushing African rebellions was an expensive business, and British taxpayers jibbed at shouldering the cost on behalf of what was seen as a community of dissolute reactionaries. Exsanguinated by the Second World War, Britain was divesting itself of a demanding empire. Why should Kenya be an exception? Accepting the inevitable, the British government invited the colony's emerging black leaders to a series of conferences in London's Lancaster House in 1960, and the shape of future self-government was gradually agreed. Three years later, after his KANU party claimed an effortless election victory, Kenyatta became the first prime minister of independent Kenya. At a ceremony in a Nairobi stadium, the British flag was lowered in tactful darkness and a new Kenyan one – black for the people, red for the blood that had been shed, green for the land – was raised to cries of '*Uhuru!*' (Freedom). Prince Philip turned to Kenyatta, seeing a lifetime's ambition fulfilled, and joshed: 'Are you sure you want to go through with this?'

What happened next underlined how thoroughly the colonial authorities had misunderstood Kenyatta. Shaking their heads at the chaos that must surely come with black rule, many settlers pocketed British government compensation, sold their farms and returned to the motherland. The exodus threatened to destabilise the economy. One of the first actions of the man the *Daily Telegraph* had labelled 'a small-scale African Hitler' was to gather four hundred nervous settlers in a town hall in Nakuru to hear a message of reconciliation. 'There is no society of angels, black, brown or white,' he told them. 'If I have done a mistake to you, it is for you to forgive me. If you have done a mistake to me, it is for me to forgive you.' In return they roared a grateful '*Harambee!*' – 'Let's work together' – Kenyatta's battle cry.

Claiming that 'We all fought for Uhuru,' Kenyatta blithely rewrote history, recasting the anti-colonial struggle as something that stretched far beyond the Kikuyu, a blurry, noble joint effort that somehow embraced black and Asian, collaborators and forest fighters, Kikuyu and non-Kikuyu. His message to demobilising Mau Mau expecting radical reform was so severe it amounted to a repudiation. 'We shall not allow hooligans to rule Kenya . . . Mau Mau was a disease which has been eradicated, and must never be remembered again.' The revolution would not take place; Kenyatta stood for continuity, not change. Executing the same nifty manoeuvre Stalin had carried out with Lenin and Mobutu would perform with Lumumba, he claimed the mantle of the great national iconoclasts even as he neutralised their legacy. The ragged Mau Mau fighters who emerged from the bush only to find both their lands and wives appropriated in their absence were swiftly marginalised. Kenyatta invited them to independence celebrations, and fawned over them in public, but the kitchen cabinet he pulled together in 1962–63 contained not a single member of the movement. Since the Kikuyu who could afford to buy the farms of departing settlers were almost always loyalists, the rich elite that emerged was solidly Home Guard.

The impact of Kenyatta's 'Forgive and Forget' slogan – historians refer to a policy of 'orderly amnesia', of 'therapeutic forgetting'[23] – linger to this day. 'We don't care where we're coming from, we care where we're going to,' a Kikuyu will tell you in justification, but the relationship with the past is more complicated and tortured than that. In the public consciousness, a hypocritical history has taken convenient root. Just as it is sometimes impossible to find a Briton who voted for Margaret Thatcher, and every Frenchman's father appears to have been in the Resistance, every Kikuyu seems to have had a father who fought valiantly for Mau Mau. When the NARC government, which rescinded a colonial-era ban on Mau Mau that remained on the statute books, unveiled a statue to Mau Mau leader Dedan Kimathi in central Nairobi in 2007, no mention was made at the ceremony of the existence of a pro-settler loyalist movement. Significantly, recent books on Mau Mau and the colonial era in Kenya have all been written by white Westerners.[24] 'When it comes to Mau Mau, a terrible pall of silence hangs over Kenyan intellectual life,' says John Lonsdale, a Cambridge professor who has dedicated his career to shattering that taboo. 'Kenyans may write their autobiographies, or record the pre-colonial histories of their ethnic communities. But they don't write about Mau Mau.'

Elderly Kikuyu living on what used to be the Kikuyu Reserve but is now just Central Province, a striking number of whom still suffer physical side-effects from being beaten with rifle butts by British soldiers or held too long in handcuffs – a stiff hip, a faltering walk, annoyingly nerveless fingers – retain a mental map of the landscape shaped by Mau Mau. These eighty- and ninety-year-olds can show visitors the location of the caves where fighters hid and were smuggled food, the spots where the Kikuyu were herded into artificial villages, the junctions where the disembowelled bodies of vanquished Mau Mau – their intestines wrapped around their torsos like bandoliers – were displayed. But none of these features on any map or in any tourist guidebook, and this silent topography will gradually disappear from community consciousness as the elders die.

In Nyeri, a cement obelisk on the main shopping street supposedly pays tribute to Mau Mau's fallen, but the plaque explaining this is missing. As pedestrians bustle past, it sits blank, ignored, anonymous. Perhaps the most creepily poignant site lies at the gravel entrance to the town's golf club. Some fifteen years ago, the story goes, workmen were sent to fill a dip that kept forming under the chairman's parking space after each heavy rain. They began digging, but dropped their tools in alarm when smoke began mysteriously wafting from the open ditch. The neat greens are located, as it happens, on the site of a former British prison, and today's parking lot lies where the bodies of hanged Mau Mau were thrown. Rationalists may reject the tale as an urban legend, but the story certainly contains a metaphorical truth. In local minds the Mau Mau era, like the unrecorded bodies of its dead, continues to fester underground like so much toxic waste, ready to rise up and overwhelm today's Kenyans with its noxious fumes.

Of course there were grumbles amongst the Kikuyu at Kenyatta's snubbing of Mau Mau. But the awkward fact that it was the collaborators, rather than the heroes of the revolution, who inherited the earth in independent Kenya was pushed to one side as the realisation set in that there was serious money to be made. Hundreds of new schools, roads and hospitals were being built, thousands of jobs once available only to whites and Asians were opening up in the state sector, and prices for tea and coffee – which the Kikuyu were now free to grow – were high. This was when the Kikuyu determination to embrace the white man's ways really paid off.

Kenyatta had revealed the expansionist plans he nursed for his community during the Lancaster House Conferences, to the dismay of other delegates. 'He said that the Gikuyu must be allowed to take up land in the Rift Valley . . . Immediately there was a long-drawn-out "Aaah" from the Kalenjin and Maasai representatives, and Willie Murgor from the Eldoret area produced a whistle and blew a long note of alarm on it,' recalled Michael Blundell in his memoirs.[25] Borrowing money from Kikuyu banks and Kikuyu businessmen, tapping into the expertise of Kikuyu lawyers, the president's fellow tribespeople rushed to buy the land of departing whites under a million-acre resettlement scheme subsidised by London. Descending from the escarpment, they flooded in their hundreds of thousands into the previously off-limits Rift Valley, seizing lands the Kalenjin and other communities regarded as having been temporarily appropriated by the white man, but rightfully theirs. Given a selling scheme based on the principle of willing buyer, willing seller, there was little the poorer tribes could do.

The Kikuyu knew in their hearts that they were doing unfairly well out of the Kenyatta presidency. But those fortune favours can always convince themselves their luck is somehow deserved. It was *their* community that had suffered at the hands of the British, the Kikuyu told themselves, *their* community that had risen up against the oppressor, *their* community – better-educated thanks to its early exposure to the missionaries – which taught less politically-aware Kenyans what it meant to be free. More sophisticated, cannier than their fellow Kenyans, they had led the way in these, as so many other areas, and had surely won in the process the right to both lead the country and eat their fill. By 1971, the conviction that this pleasant state of affairs should be rendered permanent had so hardened in Central Province that a party within a party was formed – the Gikuyu, Embu, Meru Association (GEMA), whose aim was to change a constitution which provided for vice president Daniel arap Moi, from a small coalition of Rift Valley pastoralists known as the Kalenjin, to take over in the event of the president's death. If ever there was an expression of ethnic hubris, GEMA was it.

That golden era ended in 1978, when Kenyatta took ill on a podium in Mombasa, collapsed in the men's toilets and later died. Despite GEMA's best efforts, the presidency went to Moi, who could now take his revenge after years of being patronised by Kenyatta's Kikuyu cronies. His power would be built on Kenya's smaller tribes' fear of a repetition of Kikuyu rule. Moi's publicly declared philosophy might be '*Nyayo*' – to walk in the 'Footsteps' of the revered Kenyatta – but for the Kikuyu, nothing would be the same again. It was now the Kalenjins' turn to 'eat' at the trough of the state. The Kikuyu still flourished, but they now did so in spite of government patronage, rather than because of it. In Nairobi, the *matatu* routes, the taxi trade, the hotel business, real estate – areas where the domineering KANU government enjoyed no control – were all in Kikuyu hands. GEMA went into voluntary liquidation in 1980, its dreams shattered.

Once Moi gave in to pressure to end single-party rule in 1991, it was natural that the discontented Kikuyu community, at the forefront of every curve, should launch the first opposition parties.

Kikuyus today cite this as evidence that not only were they responsible for Kenya's first liberation – from colonial rule – they should also be thanked for its second, from the one-party system. In every election that followed, Nairobi and Central Province would repeatedly, fruitlessly, vote against KANU, a constant reminder to Moi that this important section of the community rejected what he stood for. When the first serious ethnic violence in Kenyan history broke out in the early 1990s, with 1,500 'foreigners' who had settled the Rift Valley during the Kenyatta years killed by local Maasai and Kalenjin and hundreds of thousands brutally cleansed – with the support of the police and government officials – the Kikuyu interpreted it as a warning that ethnic extermination was not entirely out of the question. 'Lie low like envelopes or be cut down to size,' declared Moi's chauvinistic Maasai minister for local government, William Ntimama. Whatever protestations Moi made that he was Father to One Nation, the Kikuyu would see this bloodletting, an early signal of what the future held that no one wanted to heed, as punishment for a successful community's defiance.

It was sometimes hard to tell exactly where government incompetence ended and deliberate sabotage began. But the collapse of the coffee industry, troubles in the tea factories, the decline of Kenya Cooperative Creameries – all involving sectors at the heart of the rural Kikuyu economy – would be viewed by the Kikuyu as part of a malevolent plot to pauperise the tribe Moi feared. And they pointed to the state of the roads, schools and hospitals in Central Province as further proof of the president's vindictive determination to make them pay for past 'eating'. While Eldoret, Moi's home town, got what every analyst agreed was a superfluous airport and bullet factory, the Kikuyu got potholes and schools more like farmyard barns than educational facilities. That might not have been so bad if the country as a whole was prospering, the thinking went, but just look at Moi's pathetic economic record and compare it with the growth rates of the Kenyatta era. This was what you got when a bunch of illiterate herdsmen were allowed to run the country.

* * *

Such, then, was the community from which John Githongo hailed. He was a member of the House of Mumbi, a house whose story was in many ways synonymous with that of Kenya itself. It was a community which had a ragged wound running through it – that Mau Mau versus Home Guard rupture – and was divided and self-critical to a degree rare in other tribes. But when it turned its face to the outside world, it managed to combine a bitter sense of grievance with a superiority complex nurtured during the long years of Kenyatta indulgence.

Most African countries have their version of the Kikuyu: hard-working, economically aggressive ethnic groups whose success in business, skill at interacting with the globalised economy and bumptious faith in their own prowess so intimidate the rest that the fear shapes a nation's destiny, reducing politics to a none-too-subtle expression of resentment by the less successful. In the Democratic Republic of Congo it is the Luba, in Nigeria the Ibos, in Rwanda the Tutsi, in Cameroon the Bamileke, in Ethiopia it was once the Eritreans. The 'Jews of Africa', these groups often dub themselves, and the things once said in Europe about the Jews are muttered about them: 'All they care about is money, money, money,' 'Give one a job and the whole clan takes over,' 'They keep themselves to themselves, just can't be trusted.' And when things turn nasty, and politicians whip up ethnic hatred to please the crowds, it is these groups that pay the price.

One of the characteristics the British left behind in Kenya was a very Anglo-Saxon enjoyment of jokes. Kikuyu jokes are legion, as often as not cracked by '*Kyuks*' themselves, who have reclaimed their derisive nickname with the same confidence with which they once reclaimed their land.

Two newborn babies are lying in the maternity ward, and careless nurses get them mixed up. How to establish which is which before the very Luo Mrs Otieno and the very Kikuyu Mrs Kamau come to pick them up? 'Easy,' says Matron. 'Just jingle some coins in front of each and see what happens.' One baby falls asleep. The other wakes and holds out a pudgy hand. 'See?' says Matron. 'That one's Otieno, that one Kamau, end of story' . . . How can you tell if a Kikuyu is dead or only faking? Drop your wallet next to his bed, and if he doesn't immediately reach for it, he's definitely for the morgue . . . Then there's the one about the Kikuyu conductor of a crashed *matatu* who complains that his passengers keep dying before paying for their seats; and the one about the Kikuyu suitor whose idea of a romantic first date is to give the girl a hoe, take her to his *shamba* and put her to work.

The Kikuyu, in the popular mind, account for both the best and the worst in Kenyan culture. On the one hand, the vast majority of the ambitious youngsters who join the African diaspora each year, heading to the United States, Britain and Canada in search of business degrees and professional training, are Kikuyu. On the other, they make up a bigger share of the Kenyan prison community than any other ethnic group. 'Where you find Kikuyu, there you find thieves,' goes the saying.

When I asked an urbane Kikuyu banker, John Ngumi, for a summary of Kikuyu qualities, he produced, at machine-gun speed, a list of characteristics which juxtaposed unabashed ethnic pride with a clear insight into just why so many of his fellow countrymen found the Kikuyu irritating. 'We're thrusting, we're loud, we're hard-headed and we're everywhere. We're too many, we're greedy, many of us lack finesse, even our table manners leave a lot to be desired. We're Africans in the raw, we don't make apologies for what we are. We're the ones who keep Nairobi fed and watered and provide a host of small services that keep the country running. The problem is there aren't enough of us to dominate, yet we're too large to ignore. We are at once both obnoxious and indispensable.'

Travelling Kikuyuland, a common note emerges, whether one is interviewing a snaggle-toothed ninety-year-old farmer in the Kiambu hills or an elegantly suited banker in a Nairobi bar, and it is one that swiftly becomes wearing. It's the note of entitlement: a sense of being special, different, better – and therefore more deserving. 'If you did an experiment and took five Luos, five Luhyas, five Kambas and five Kikuyus and gave them the same amount of money to invest, the

Kikuyu would be far, far ahead,' a Kikuyu businessman will say without embarrassment. 'We simply work harder than other Kenyans.'

The great irony is that over twenty-four years of Moi rule, a community viewed as a national bully by the nearly 80 per of Kenyans who were *not* Kikuyu would come to think of itself as put-upon victim. For despite all Moi's efforts to redress the post-Kenyatta dispensation, Central Province and Nairobi, a heavily Kikuyu city, still do better than any other region of the country.[26] 'The notion that the Kikuyu have been marginalised is total baloney, a persecution complex,' says economist David Ndii, himself a Kikuyu. 'If you look at any indicators – health, education, access to piped water, access to electricity – the Kikuyu always come out on top.' He is equally scathing about the notion that the community contributes more than its fair share economically. 'When you examine the data it turns out to be bullshit. The most productive people in Kenya, in terms of output per head, are pastoralists. One pastoralist can own two hundred head of cattle. The Kikuyu don't understand geography and they don't understand mathematics. These myths have to be exploded.'

Perhaps that sense of superiority is rooted in the peasant's wonderment at being blessed, in a largely arid country, with the miracle of moist and fertile soil. Spooning up *githeri*, a mixture of beans and maize kernels, on the *shambas*, residents will cite Isaiah 18, which speaks of a land beyond the 'rivers of the Cush', and a people 'tall and smooth-skinned . . . feared far and wide, an aggressive nation of strange speech, whose land is divided by rivers', as proof that the Promised Land, far from being located in Israel, is actually to be found between the blue foothills of Mount Kenya and the misty mountain ranges of the Aberdares.

The men around Kibaki assumed that, as a Kikuyu, John Githongo would share that sense of entitlement. He would know that certain matters were best kept from those outside the House of Mumbi. Secretiveness had always come naturally to the Kikuyu, takers of terrible oaths: it was one of the traits the British had singled out for criticism, and the lean Moi years, when ethnic solidarity was the only defence possible against a rigged system, had underlined the importance of discretion. 'Matters of the heart are not like the palm that greets everyone,' goes a Kikuyu saying.

In addition, they knew that John would have imbibed a deep respect for his elders as part of his upbringing. In rural Kikuyu villages it was unheard of for a child to criticise or interrupt an elder. Directly addressing a parent was regarded as unacceptably brazen – 'How are things on the mother's side?' was the polite, if roundabout, way of saying, 'How are you, Mum?' – and children were expected to stand when an elder entered a room, or to step into the bushes if they met an adult on the road.

Admittedly, John's ancestors originated not from Kibaki's Nyeri but from Kiambu, an area whose inhabitants were regarded by Kikuyu further north as sneaky and deceitful. But what really mattered was that he was a Kikuyu. His father, accountant for Kenyatta, was privy to the Kikuyu elite's financial secrets. Joe Githongo had paid a personal price during the Moi era, and proved his credentials during the rumbustious multi-party years, working as a fundraiser for Kibaki's Democratic Party. The Githongo family had prayed and played with other Kikuyu families, and John had gone to school alongside the scions of the country's leading Kikuyu dynasties. If he could not be trusted to take the interests of the clan to heart, to instinctively grasp what mattered to the House of Mumbi, then who could?

At this point a question poses itself, one that may never be satisfactorily answered. When the elderly members of TI's board put John Githongo's name forward for the post of anti-corruption chief and their friends in government enthusiastically agreed, did they do so anticipating that one day they would need to appeal to his sense of tribal solidarity? Was his appointment, which originally seemed so well-intentioned, in fact the most cynical of political moves, the propelling of an impressionable young man into a position where, should a crisis develop involving his own community, he would find it virtually impossible to resist outside pressure? Did they name him intending to compromise him?

John certainly tends to that view. His feelings towards members of the group, with the exception of Harris Mule, are far more bitter than those towards any players in the Anglo Leasing affair. In his view, the old men he had trusted with his fate behaved like Abraham preparing his son Isaac for sacrifice. 'The *wazee* . . .' He shakes his head in wonder. 'They set me up. I was the puppet, and they the puppeteers.'

My own suspicion is that they possessed no such clarity of vision. They chose John chiefly because he was the obvious candidate. The instinct to entrust a potentially sensitive post to a fellow tribesman was certainly there, but it was not particular to this clique or period. It is the bane of Kenyan life, skewing employment patterns in every sector of the economy. At the back of their minds, the old men may have vaguely sensed that having 'one of ours' in this key post might one day prove helpful. 'The assumption must have been, "If he gets out of line, his father will have a quiet word,"' guesses Wycliffe Muga, columnist for the *Daily Nation*. 'But when you feel someone is part of what you are, you simply take it for granted he will go along with what you do.' Flush with post-election ambition, incapable of imagining a time when things were not going their way, they were more slapdash than cynical in their strategising.

In the film *The Godfather* there comes a moment when, with his father lying wounded in hospital and trusted allies being picked off by a rival mob, Al Pacino's Michael Corleone must decide whether to rally behind the clan or walk away. The former black sheep, who until that moment has shown nothing but disdain for his Mafia heritage, never wavers. Without waiting to be asked, he jettisons a lifetime of liberal values, personally executes his father's enemies and takes charge of the family, going on to become the most ruthless godfather of all. If they ever bothered to think about it, the Mount Kenya Mafia must have assumed John Githongo would do likewise. Oh, he might squawk and fret a bit, given his university education and professional credentials, just to show he was a man. But he would simply know, without having to be told, exactly where his duty lay.

8

Breaking the Mould

'The hope of a secure and liveable world lies with disciplined nonconformists.'

MARTIN LUTHER KING

In assuming their secrets were safe with a reliable insider, the ruling elite had simply not done its research. To John Githongo, being Kenyan, being Kikuyu, did not mean what it meant to his colleagues. If they were not aware of that awkward fact, he certainly was.

The Githongo family home is in Karen, a suburb twenty kilometres south-west of Nairobi which looks across the knuckled ridge of the Ngong Hills, formed – legend has it – when a giant who had tripped over Mount Kilimanjaro pressed a fist into the ground to right himself. The area lay a full afternoon's ride outside the capital when Danish baroness and writer Karen Blixen, to whom the suburb's name is mistakenly attributed,* tried and failed to make a go of her coffee farm in the 1920s. Nowadays, forest still separates Karen from the city, a stretch of woodland crossed at speed at night by nervous drivers braced for car-jackings and hold-ups. But the spate of new building is so rapid, with each week seeing another expanse of

* It is actually named after Karen Melchior, Karen Blixen's cousin, whose father owned the Karen Coffee Company.

yellow scrubland transformed into a construction site, that soon the trees will be gone, the stables supplying the glossy horses clip-clopping along the lanes will close, and Nairobi will have pulled Karen permanently into its cement embrace. In colonial days, this used to be a white area, and many of the name plates peppering the high bougainvillea hedges are as British as empire itself: Hardy, Roberts, Hughes, McRae. This is still a haven for the breed known locally as the 'KCs' – Kenya Cowboys – rakish descendants of Kenya's original white settlers, who are proud of their fluent Kiswahili, refer to their staff as 'the *watu*', drive their Land Rovers into town as rarely as possible and try to ignore the urban development creeping steadily towards them. But colonialism's remnants are now outnumbered. The black elite long ago made Karen its own, and the faces on the eighteen-hole golf course of the Karen Country Club these days are overwhelmingly brown.

The Githongo house is an old colonial villa, red-tiled, solid, timber-framed. At six acres, its grounds are large enough to hold outhouses at the front and a small livestock farm at the back. The Githongos may belong to Kenya's aristocracy, but the house feels more like a working homestead than a showpiece residence. It has the utilitarian feel of a family home that sees a regular throughput of children and grandchildren, colleagues and in-laws, girlfriends and schoolmates, servants and hangers-on. In the front compound several ageing cars await a mechanic who may never arrive; the veranda overlooking the lawn is used for hanging laundry, not serving gin-and-tonics, and the swimming pool was drained long ago.

The paraphernalia on family mantelpieces usually reveals what people regard as important in their lives. On this one, below David Shepherd's classic painting of a bull elephant, sits a framed photo of the three Githongo boys, taken, to judge from the flared white suits, puppy fat and semi-Afro hairstyles, sometime in the mid-1980s. Nearby perch two snaps showing John's parents shaking hands at the Vatican with Pope John Paul II. The religious theme is even stronger in the dining room. Portraits of Kibaki and Kenyatta – no Moi – compete with images of a doleful Messiah, bleeding heart exposed.

'There is surely a future and hope for you and your hope will not be cut off,' reads the message on one blue-tiled plaque. 'CHRIST is the HEAD of this house, THE UNSEEN GUEST at every meal, THE SILENT LISTENER to every conversation' warns another, with a touch of menace. This, quite clearly, is a household in which fork is never lifted to mouth without grace first being recited and chests respectfully crossed.

Joe and Mary Githongo have lived here for thirty years. Theirs was a relationship forged against the odds, for it straddled the divide that fractured Kikuyu society in the 1950s. Both came from Kiambu, the province abutting Nairobi. But their families were from opposite sides, in more ways than one.

Joe was born a Catholic, his father a well-travelled truck driver and cattle rearer prominent in the Mau Mau. A member of the Ndege age-set that formed Mau Mau's ideological core, John's paternal grandfather was a hardliner who administered the dreaded oath, raised funds for the movement and was arrested by the British and beaten more than once. In old age, he would recount with chortling relish his role in the garrotting of a loyalist chief. 'He was right in it, very deep,' remembers his daughter-in-law.

Mary, in contrast, was a Presbyterian and her father was a member of the Home Guard, his loyalist credentials impressive enough to win him a senior chief's appointment from the British. Growing up in a guarded village, complete with spiked moat, watchtower and armed sentries scouring the horizon for 'terrorists', she remembers the tension when night fell, bringing with it the fear of attack by the men in dreadlocks, seen not as heroes but as bloodthirsty marauders. She also remembers being jeered at when she went to the river to wash clothes by children whose parents were Mau Mau. 'I was too young to know what they meant, but later I understood. If you belonged to a person like my father, no one would like you.' Today, she believes her father hedged his bets by taking the oath in secret – many Home Guards did exactly that – and that he was 'playing double'. But in the absence of any evidence – and Mary Githongo admits she never discussed the matter with him – it's hard to see this as much more than a wistful attempt to paper over the awkward fact that one side of her future family once regarded the other as colonial toadies of the most loathsome kind.

As the son of a suspected terrorist, Joe Githongo did not escape the Emergency unscathed. Aged twenty-five, an employee for Kenya's Post Office, he was arrested, categorised as a 'grey' detainee (not entirely trustworthy, but not a hardliner either) and held for three months at the infamous Manyani Reception Centre – 'a terrible, terrible place'. The authorities' assumption that he had taken the oath was not misplaced. 'I had taken it because I believed we had to fight for our own government.' That opinion was put to the test as camp organisers set about destroying Mau Mau's hold on Kikuyu inmates. White missionaries were called in to reverse what was seen as Mau Mau brainwashing, telling the prisoners they could still be saved if they admitted what they had done. As a devout Catholic who nonetheless believed in home rule, Joe was pulled in opposing directions: 'It was a time of terrible turmoil.'

Today, his memory of this period is clear-eyed. He retains a vivid mental image of one white sadist, a prison officer. 'I would watch him from my office in Gilgil. He would stop by at eleven in the morning to chase a Maasai herdsman. He would stop his car and beat the man, for nothing, even if it meant soiling his nice khaki uniform. He loved it. I would watch and think: what is going on with this educated man?' But Joe also singles out for praise 'remarkable, outstanding' British district officers who stood up for ordinary Kenyans while the Mau Mau were doing 'unspeakable' things. 'During Mau Mau I saw the worst in the Kikuyus and the worst in the British. There were terrible excesses on both sides. Our politicians went too far. I approved the noble objectives at the beginning. But we didn't need to do all that killing.'

Little wonder that after being released for good behaviour, Joe felt an urge to leave the country. In 1959 he headed west, to the Congo's Leopoldville, where he eventually got a job as an accountant for the United Nations force called in to stabilise a country in a state of post-independence ferment. After three years banking his UN salary, he moved to Hampstead in north London, whence he summoned the young woman he had wooed back in Kenya. For Mary, the decision to marry a Catholic, a member of a religion with which she found herself increasingly in tune, meant a break with her Protestant family. 'When I first told my father I wanted to change my religion he said: "You leave this place." So I waited until I got married and then I became a Catholic. The Lord just called me.'

Modern-day Kenya is full of couples whose parents, had they encountered each other during the Mau Mau years, would gladly have killed one another – such is the nature of civil war. The Githongos were no rarity in this regard. But it's easy to see how that divide could serve as the first in a series of small alienations weaning a young couple away from its roots. However tough life abroad might prove, it would always be easier to navigate than home, place of unacknowledged, bitter grudges and recent, incestuous violence.

Finding a job in 1960s London was not easy for Joe. 'I remember turning up two, three, four times and being told "It's already taken" because I was black,' he recalls. He finally became a British Rail guard at King's Cross station, an undemanding post that allowed him to study for 'A' Levels while on duty. Mary worked as a nurse for the National Health Service. John Githongo Muiruri, the couple's first child, was born in London, premature and worryingly small. The future Big Man started out as the runt of the family. Siblings Gitau and Ciru swiftly followed. John was three when Mary Githongo, finding the task of rearing a young family single-handed in London exhausting, insisted on returning to Kenya.

There was never any question of the family returning to Joe's homestead near Mangu, seventy-three kilometres north-east of Nairobi. For a couple with their experience and Joe's newly acquired diplomas, Nairobi was the place. Having left behind a country under colonial rule, they were returning to an independent African republic, where so many things once ruled off-limits for blacks had become possible. Lowly public-sector workers in Britain, they belonged to a fast-emerging middle class in Kenya, a class that had signed up for Kenyatta's nation-building project and had no desire to look back.

Joe joined the management of Kenya Railways, while building a client base for the accountancy firm he planned. By 1969 he was ready to launch it: Githongo & Company, one of Kenya's first black accountancy firms. That alone ensured it a huge amount of work, as the Kenyatta government channelled auditing contracts for Kenya's state-owned concerns towards a company seen as being 'one of ours'.

The role of accountant is a little like that of priest or doctor. Whether he wants to or not, the practitioner acquires an intimate knowledge of his clients' less creditable affairs. Large, rotund – the Githongos don't *do* small – scrupulously polite, Joe inspired the necessary trust amongst those with plenty to hide. He acquired a reputation as a man who, when sudden misfortune fell, could be relied upon to appear with a briefcase of cash to magic the problem away. With the presidential family as its most high-profile client, Githongo & Company handled the affairs of a large chunk of the post-independence elite. Joe was appointed secretary to GEMA, and many of the Kikuyu entrepreneurs at the heart of that movement passed through his doors. 'Everyone who was anyone was audited by my dad's firm,' recalls Mugo, John's youngest brother. 'Our growth,' remembers Joe with fond nostalgia, 'was phenomenal.'

Then Kenyatta died, and the firm's fortunes turned. 'When Moi saw Kikuyus doing well, he would intervene. He actually openly attacked our company in Nyeri in 1984 when he discovered we represented a lot of the tea factories in the country, saying some of those jobs should be taken from us,' remembers Joe. Audits for state enterprises were systematically channelled elsewhere. 'We lost more and more of these jobs and had to take on more and more work with private companies.' The impact on the family was swiftly felt. Mugo is the only Githongo child without a university degree. He was forced to cut short his education in his late teens, when money suddenly got tight.

It was ironic that Githongo & Company should fall foul of the Moi regime's ethnic hostility. For the one characteristic acquaintances and friends all remark upon when discussing Joe Githongo is his loathing of tribalism. Perhaps his visceral horror at the excesses of the Emergency, when ethnic chauvinism devastated his community, explains the sudden force that comes into his voice when quizzed on the subject. 'I hope people come to associate tribalism with corruption and throw it out completely.' Many Kenyans of his generation would ostensibly agree with such sentiments, without acting upon them. Joe put his principles into practice, scrupulously appointing members of other tribes to the board of Githongo & Company to dilute the Kikuyu quotient.

At home, the five Githongo children – the last two were born in Nairobi – were brought up to be proud of their nationality, not their ethnicity. 'The Kikuyu thing is of no consequence. My children were brought up as Kenyans, not as Kikuyu,' insists Joe. He tells with pride the story of how when John visited friends in western Kenya, his dark skin led many to assume he was a local. A more hidebound Kikuyu might have found the idea offensive. 'John was happy to be an honorary Luo.' Neither parent would object when their children befriended, dated and eventually, in some cases, married non-Kikuyus.

The cosmopolitan message was reinforced by John's secondary school, St Mary's. Founded in 1939, this private school was one of several established in Kenya for white settlers determined that their children should receive an education on a par with anything available back home. With the passage of the years, the white boys were replaced by the children of Kenyan permanent secretaries, special branch chiefs, army generals and government ministers. For a *nouveau riche* who had only just knocked the clinging red mud of Central Province from his shoes, schools like St Mary's were an aspirational dream, offering legitimacy in one swift generation.

Located in the well-heeled suburb of Lavington, built on the site of Kenya's first coffee farm, St Mary's – or 'Saints' as it was nicknamed – was structured on lines any fan of Harry Potter's Hogwarts would recognise, so bent on reproducing the traditional British public school experience the effort verged on parody. There were prefects and head boys, uniforms (blue blazer and tie) and houses, a Latin school motto (*Bonitas, Disciplina, Scientia* – 'Goodness, Discipline,

Knowledge') and assemblies, and the year was divided into Michaelmas, Lent and Trinity terms. Set up by members of the Holy Ghost Fathers, an Irish order, with a full-sized Catholic chapel on the premises, the school's emphasis was spiritual and religious, even if a certain leeway was permitted on the question of precisely which faith was involved. 'We accept Sikhs, Protestants, Hindus. The only thing we don't allow is atheists,' explains headmaster Henry Shihemi.

While state-funded schools like Alliance or Starehe Boys were rigorously meritocratic, the demands placed on St Mary's boys were simpler. Mr Shihemi gives me a slightly sardonic look, rubbing finger and thumb together, when I enquire about entry criteria. The school lies at the pricier end of the independent market, and does not offer scholarships. In return for the hefty fees, Saints pupils enjoyed facilities that would leave even Western children blinking in envy: how many schools boast a golf course on the premises, can muster a full orchestra, or routinely lay on end-of-term musicals and performances of Gilbert and Sullivan operettas? While state schools in Kenya were mad about soccer, the great equaliser, St Mary's prized rugby, the upper-class ball game.

The Irish, Indian, Goan and African priests who taught there drummed the notion of public service into their pupils. But it was made clear that the people who would actually lead the country in future, as opposed to dutifully ensuring its efficient running, would come from schools like Alliance. St Mary's was the Ampleforth of Kenya, and boys from the state schools did not hesitate to rub home the difference when coming into contact with Saints pupils who, for their sins, even spoke with a distinctive, cut-glass accent. 'When we played them at rugby, I'd go out of my way to hurt the St Mary's boys,' remembers writer and analyst Martin Kimani. 'In our eyes, those who hadn't made the grade but had money went to St Mary's. To us they were soft, spoiled, privileged. It was ironic that St Mary's taught the importance of public service, because the rest of us looked at them as the kids of crooks.'

The priests' lessons, it seems, ended up being digested in radically different ways. It is one of the curiosities of St Mary's that while the school – whose alumni quip 'Once a Saint, always a Saint' – produced the most famous government whistleblower in Kenyan history, it also simultaneously produced its most notorious wheeler-dealers: Gideon Moi, who built a business empire on his father's name; Jimmy Wanjigi, one of the businessmen at the heart of Anglo Leasing; and Alfred Getonga, State House insider, were all contemporaries of John's.

In the latter years of the Moi regime, when even education would become embroiled in the country's tribal tussle, St Mary's came to be labelled a 'Kikuyu school'. Leafing through the annuals from the late 1970s and 1980s, prominent Kikuyu names leap from the squash and hockey team lists. Eric Wainaina, pop star and human rights campaigner; Jeff Koinange, former CNN anchorman; David Kibaki and Uhuru Kenyatta, the sons of two Kenyan presidents; Raymond Matiba, son of the opposition leader: Kikuyus all. But that's not the way it felt to former pupils, who remember schoolmates from western, eastern and coastal Kenya, and making friends with pupils from Uganda and Tanzania, Denmark and the Philippines.

In fact, the education offered at this private school was almost quixotically international in its outlook. 'There was a strong Western tinge to it,' recalls John Gethi, one of John Githongo's old schoolfriends. 'But of course, when you were there you thought everywhere was like that.' English and Kiswahili were spoken, but Luo, Kikuyu or Kamba were frowned upon. 'We don't encourage groups to form on ethnic basis, and we don't encourage pupils to speak in the vernacular. We are proud to say we are citizens of the world,' says Mr Shihemi. The school authorities regarded education as part of a nation-building project which would ultimately do away with tribal differences. 'We will eventually reach a time where we will no longer say "This is a Kikuyu or a Luo." If more schools were like St Mary's, it might happen faster.' Friends would later trace John's extraordinary ability to mix with people from radically different *milieux* to his school's heterogeneous approach. 'We were a generation that didn't know about tribes,' recalls Gethi. 'I don't think we'll ever get back to that. It was a golden era.'

In the process of opening up to the world, certain cultural baggage had to be jettisoned. In Kikuyu tradition, circumcision looms large. The Githongo boys certainly went under the knife, but for them a rite once staged on the banks of a river, with only cold water to numb the pain, was performed in hospital. Ask John and his contemporaries about traditional Kikuyu concepts such as *wiathi* – becoming master of one's destiny – and like many urban Kikuyu their age, they will hesitate, shrug and look awkward. They prefer a Robert Ludlum thriller to the latest academic work on Mau Mau, and while they might visit '*shags*' – slang for the village ('*gicagi*') – at intervals and hold their grandparents in tender affection, that doesn't mean they know a great deal about their roots.

As the eldest son, John carried huge expectations on his shoulders. First-born children always face a tougher battle carving out their territory than children who come later. Role model to his younger siblings, more was always expected of him, less forgiven. But along with those responsibilities went an awareness of the princely importance a son enjoys in a patriarchal society. Under Kikuyu tradition, the husband commands a family's respect, then the first-born son, and only then the mother. From the first breath he drew as a premature baby, with all the extra concern that vulnerability entailed, John knew he was special.

Family opportunities came his way first, and they were always of a kind to expand a lively boy's horizons beyond Kenya's borders. In 1978, when he was only twelve, American friends of the family invited him to spend a month in Chicago. At that age, a month seems like an eternity. The trip gave John his first taste of independence. It also alerted him to the reality of racial segregation. The blacks he met in Chicago found the fact that he was staying with a white family bizarre and inappropriate, and told him so.

Aware of how his experiences abroad had changed his vision of the world, Joe Githongo set about passing those insights on to his children. On foreign trips, he would go to a video store and buy up the best Western movies of the day. 'These were films that weren't available in Kenya at the time. We would stay up till two in the morning, watching one after another,' remembers John's brother Gitau. Joe's patronage of a school back in Mangu had a similarly stimulating effect. He set up an exchange system with Western students interested in serving as teaching assistants. Each year a new youngster would arrive, to be welcomed into the Githongo household during vacation time. 'They were slightly older than us – seventeen-to-twenty-year-olds – and we would quiz them about the world. One wanted to join the SAS, and it was the time of the Falklands war – that caused a lot of family discussion. Another set off for Khartoum, was caught up in the Sudanese civil war, and returned full of stories. They were all fascinating people and I think that had a very profound influence on us all,' remembers Gitau.

It's easy to forget the sheer dreariness of Moi's Kenya, when the system sought to crush the debate that leads to political challenge. The country's small group of intellectuals felt besieged. 'Books were incredibly important, things to be cherished,' remembers Rasna Warah, a columnist for the *Nation*. Nairobi bookshops would not officially stock works deemed to offend the presidency, but they could usually be bought discreetly under the counter, hidden amongst a spray of magazines, if you knew the right code word. 'You never knew for sure what had been officially gazetted as a banned book,' says Warah, 'so you stashed your entire library under your bed. If you owned a book that might have been banned, you photocopied it and it circulated in A4 form, person to person, because then it was easy to hide amongst your ordinary papers. You felt watched all the time. When you went out, you would look for the shiny shoes. They were the dead giveaway, the very shiny shoes of the National Intelligence guy who was following you. I remember ducking under the table in restaurants to check out the shoes around me.'

The country's television and radio network was firmly in state control, private FM stations a distant dream, and the bravest independent newspaper, the *Nation*, rarely penetrated remote areas. Frustrated writers like Kwamchetsi Makokha, who wrote a radio soap opera for several years, had some fun sneaking political messages into their work, in his case the saga of a humble office messenger called Lameck. 'The censors were incredibly stupid. As long as you didn't name people or institutions directly, you could get away with the most obvious analogies and withering criticisms.' That still left millions of Kenyans hungry for more information than was offered by the universally distrusted Kenya Broadcasting Corporation, which began virtually every news broadcast with a kneejerk 'His Excellency Daniel arap Moi today . . .', while feeding its audience a steady diet of lip-glossed, shoulder-padded schlock: *The Young and the Restless*, *The Bold and the Beautiful*, *The Rich Also Cry*. Those curious to know what was going on in their own country relied on the BBC's World Service – the 6.30 p.m. news in Kiswahili was a must – or, as a last resort, on the so-called Machakos Express: a week's worth of political gossip from the capital relayed in person by Nairobi workers on weekend visits upcountry. In that torpid context, few Kenyan families can have enjoyed wider vistas than the Githongo family.

John soaked it all up, then passed his enthusiasms on to his siblings. 'He was always a maverick,' remembers Gitau. 'He would pick up a book and talk about it for days. He was the first one to explain the word games in the Asterix the Gaul comics to the rest of us. He was a great fan of Richard Pryor. I remember seeing Martin Luther King on television for the first time. I was trying to work out why this guy was half singing and half praying, but John said: "That's Martin Luther King. He's incredible." He was trying to decipher these things long before us.'

At school, John swiftly excelled. Physically, he might seem ill-at-ease in his oversize body, but intellectually he was a Renaissance man in the making. History was a favourite, a precursor of a later fascination with politics. He was a talented draftsman and also loved zoology, becoming near-obsessed with the minutiae of animal behaviour. English literature brought him into contact with the works of Wole Soyinka, Chinua Achebe, Ngugi Wa Thiongo. A born storyteller, he poured out plays and short stories. And he was liked, both by the priests and the boys, as his election as one of the school's first black head boys attested. He already possessed a charisma not evident in his bespectacled photograph in the St Mary's annual, a snapshot as stolid and uninspiring as all his official portraits.

A bit of a goody-goody in the eyes of his siblings, this younger version of John Githongo was a far bossier character than the adult, who would learn the value of self-effacement. 'He would bring his friends home from school, sit at a table and just talk. That was his own space and no one could interrupt,' remembers sister Ciru. When a handful of friends at St Mary's got together to form a gang – the Bundume Boys – John fought for pole position. 'He was very forceful, overbearing. He wanted to be the boss,' remembers gang member Gethi. 'We had our "spot" on the golf course, under a tree, where we had lunch together every day. No one who valued their lives would venture in.' Despite its macho nickname – '*dume*' means 'bull' in Kiswahili – this was no band of radicals. 'We'd get together and revise like mad, really tearing the books apart.'

By his mid-teens John was, in typical adolescent fashion, searching for an identity, trying on various personalities. He flirted with Islam, buying a Muslim cap and reading the Koran with typical singleness of purpose. 'John never did anything just for the heck of it. There was always a reason,' says his sister. When that belief system failed to fit, he swung the other way, in a direction less calculated, this time, to outrage family and friends.

In upwardly mobile families, sociologists say, it is the mother who serves as moral and spiritual compass. The Githongo family was no exception. Statuesque, grave and as imposing as a granite outcrop, Mary Githongo had always played a steadying role in the partnership, slamming the ethical brakes on when Joe, the flighty, less scrupulous one, showed signs of getting swept into one of his businessmen friends' dodgier schemes. Her conversion to Catholicism had been no marital stratagem, but a life-changing event. It's virtually impossible to find an African family which doesn't take its religion – whether Christian or Muslim – extremely seriously. But even by these standards, Mrs Githongo stood out amongst her contemporaries as exceptionally devout. She observed mass every day, a Spanish priest regularly arriving at the family house in Karen to conduct the service.

And Father Alphonse Diaz was no ordinary Catholic priest. He was a member of Opus Dei.

Established in 1928 by Josemaría Escrivá, a Spanish lawyer-priest who experienced a divine revelation, Opus Dei – Latin for 'the Work of God' – holds that each individual, however seemingly humdrum or materialistic his walk of life, can be redeemed through work. 'God created man to work,' Monsignor Escrivá preached. 'Work is one of the highest human values and the way in which men contribute to the progress of society. But even more, it is a way to holiness.' The sinister presence at the heart of Dan Brown's *The Da Vinci Code*, Opus Dei has long been regarded in Europe with suspicion. Its critics denounce it as a secret society, whose furtive ways border on Christian Freemasonry. Pointing to historic links with reactionary regimes in Europe and Latin America, they accuse it of preaching a form of Catholo-fascism. The movement certainly makes no secret of its practice of targeting a society's elites for recruitment. 'If we don't start at the top we'll never reach the bottom,' one member in Nairobi unabashedly told me. 'The snow is at the top of the mountain. Unless it melts, you cannot irrigate the countryside'. The tactic has prompted accusations of infiltration plans and hidden agendas. Why so secret? Why so focused on the rich and successful?

But in Kenya, where Opus Dei first set foot in 1958, the organisation has made little effort to cover its tracks. Since the British colonial administration traditionally left education to the churches, whether the White Fathers, the Methodists or the Scottish Presbyterians, few eyebrows were raised when Opus Dei set about opening educational establishments in Nairobi. Today it runs several private schools and has won an official charter for Strathmore University, a neat grey-stone college built in the middle of a tatty housing estate. Hoping to transform Kenya by churning out generations of disciplined professionals, Opus Dei specialises in courses in accountancy, business studies, technology and management – dry disciplines suited not to the country's future political movers and shakers, but to the technocrats who help those with crowd-pulling talents realise their dreams. What more perfect fit could there be than the Githongo family?

John began attending Hodari Boys, an after-school club organised by Opus Dei. It offered trekking, cycling and camping holidays to energetic youngsters, preaching a practical message of self-discipline and moderation. For the Opus Dei clergymen who ran Hodari Boys, guzzling four cans of Coca-Cola on a hot day was no better than getting swinishly drunk: both actions showed a lack of self-control. John never became a member of Opus Dei or signed up as a 'cooperator' – one of those who help the organisation via charitable donations or prayer. But when he spoke as a guest lecturer at Strathmore University many years later, he attributed his working methods to his time at Hodari Boys, which he said had taught him how to plan and prioritise. The source of those little black notebooks in which John recorded his days, transcribed important conversations and listed tasks to be completed becomes clear. It was at this age that he set about meticulously recording the details of his existence. It was almost as though, in some strange form of predestination, he began in childhood honing the skills, the compulsive keeping of accounts, that the Anglo Leasing affair would demand of him in later life. After the years of adolescent searching, Opus Dei bestowed the sense of moral order his kaleidoscopic personality craved.

If the intensity of Mrs Githongo's religious beliefs created a divide between the family and other members of the Karen set, so did another family peculiarity: Joe did not drink. Kenyan socialising, like its British equivalent, rotates around alcohol. Joe's teetotal habits made him a bit of an odd man out, an impression furthered by his lack of interest in golf, that obsession of the Kenyan elite. Joe certainly belonged to the Karen Country Club – he could hardly afford not to, given the networking his job demanded – but he was never one to prop up the bar over a clinking collection of Tusker bottles. He had attended the same school as Kibaki, Mary had been taught at nursing training college by Kibaki's wife, and the couple's political affiliations were clear, with Joe hosting several Democratic Party fundraisers on the villa's front lawn in the run-up to the 2002 elections. But John's siblings nonetheless remember an isolationism that seemed to filter down from the head of the family. 'My dad is a single child, a loner, not a social person,' says Mugo. 'People have considered the Githongos to be snobbish.' At evenings with other successful Kikuyu families, the Githongos were always the first to leave. Inside the villa's walls, it was easy for the family to feel like a universe unto itself.

When the time came to head for university, John once again benefited from advantages his farming ancestors in rural Kiambu could barely have imagined. His parents offered to fund a course in Britain: a BA in Economics and Philosophy at the University of Wales in Swansea. The Githongos' faith in the value of a Western education was absolute, and each child would acquire at least part of their education abroad. The economics part of John's course left him sceptical, but philosophy was another matter. He gobbled up Plato and Socrates, Leibnitz and Descartes, Russell and Wittgenstein.

Those three years in Wales brought home to John, just as St Mary's had done, the realisation that the world extended far beyond Kenya's parochial horizons. But he also became aware that not everyone responded as he did to this widening of perspectives. Student life in rainy Wales could be grim and lonely. For his fellow Kenyans, amongst whom the Kikuyu predominated, homesickness brought out an ethnic chauvinism John had not glimpsed before, a nervous retreat into the bunker of the tribe. 'Being outside Kenya is the moment where you discover your African roots.' He had gone through a phase of intense interest in Kikuyu culture, dragging Mugo off to listen to 'one-man guitar' performers from Central Province, whose plaintive melodies are East Africa's version of country and western. Possessed of a keen ear, John spoke far better Gikuyu than many of his contemporaries, a Gikuyu picked up listening to his grandparents on the *shamba*. But he was too astute not to notice the negative aspects of this ethnic nostalgia. Noting the way the various ethnic cliques in the Kenyan diaspora sniped at each other, the jokes – supposedly ironic – about the need for the House of Mumbi to stick together, John pondered just how Kikuyu he felt, and whether such bonding was quite the positive experience his fellow students seemed to believe.

John returned to Kenya in 1987 to face the full onslaught of paternal expectation. Joe Githongo had never made a secret of the fact that his ultimate ambition was to turn 'Githongo & Company' into 'Githongo & Sons', and as the eldest son, the duty fell to John. He tried. But mathematics had never been his strong point. And there is no fussier taskmaster than the ageing founder of a family business, aware his powers are fading, yet constitutionally incapable of handing over the reins. 'It's a nightmare working with an entrepreneur who is past his creative peak. My father wanted to bequeath the firm to a loyal son, but once you got into the office you were expected never to disagree with him,' recalls Gitau. There were noisy clashes, quarrels, with Joe sacking both John and Gitau, only to take them on again. 'John worked for him about three years, but he hated accounts.' It was a defection that left his father bitterly wounded. In the West, a bright graduate's decision to turn his back on a family firm and try his luck in a huge jobs market causes little surprise. In Africa, where opportunities are so much more limited, the choice astounds. John's failure to respect his father's wishes constituted the first in a series of jarring acts of revolt. 'It was always clear to me he was a renegade,' a Kenyan journalist friend once commented. 'I couldn't understand why others were surprised. I mean, he refused to join the family firm, didn't he?'

It might seem quixotic that at almost the same time as he was chafing at his father's heavy paternal hand, John should briefly sign up as a police reservist, the force's youngest. But for John, it was a perfect way of satisfying his inquisitiveness about his own society. At the weekends he would borrow his mother's car, pick up two armed police officers and drive at night into parts of Nairobi where the criminal gangs had their lairs and an upper-class Kenyan like him never normally ventured. He found himself driving towards armed robberies while the bullets were still flying, chasing a runaway truck through the Nairobi traffic, picking up the human debris from drink-drive accidents and ferrying the dying to hospital and the dead to the morgue. Looking through a policeman's eyes, he became aware of a previously invisible underclass. 'I found that in Karen there were huge slums which I had never even seen. In Lenana, there are thousands of people living on a rubbish tip. I'd been there thousands of times and I'd never even seen them.' He quit when the police force started using reservists for crowd control. Beating up fellow Kenyans demonstrating for greater political openness was not what he had envisaged on joining.

Another small rebellion was John's announcement that he was leaving the family home. He rented a bachelor pad in Nairobi's Riverside Drive, where first Gitau and then Mugo joined him, the three determined to demonstrate their financial independence. In this all-male set-up, John could establish his own routine. His frenetic networking would always need to be balanced by long hours of solitude in which to muse, to read, to be himself. And it was easier, living with his brothers, to structure his day in the way that suited his decidedly unconventional body clock – working till the early hours when a job needed doing, then sleeping into the afternoon, so soundly it seemed impossible to wake him. It was an anti-social routine that drove girlfriends mad, and more than one walked out, declaring him too eccentric for her tastes.

'I don't want to live the bourgeois life,' John declared, and with his role as heir to the family business rejected, he threw himself into his own projects. He saw his future in creative writing. Along with a friend, Martin Khamala, he had already set up a company dubbed 'Mank and Tank', which produced a cartoon strip set in the offices of a multinational company. John wrote the dialogue, Khamala drew. It's easy to imagine how Mr and Mrs Githongo must have felt about their first son, prodigy, star pupil and former head boy, wasting his energies on a lowly comic strip. But then, Mr and Mrs Githongo were not consulted. The strip came to the attention of *Executive*, a glossy business magazine which had won itself a reputation as a cauldron of progressive opinion. *Executive* commissioned Ali Zaidi, an Indian subeditor, to run a feature on the duo. John's first appearance in the Kenyan press would be as 'John Githanga' – Ali got the spelling wrong.

Ali Zaidi presides over Nairobi's version of a French *salon*. Anyone with something of interest to say, preferably carrying a bottle or two, is welcome to sit on his veranda and meet whoever else has decided to pop round. He bonded with John, getting drunk with him, swapping books and ideas. It was Ali who gave John his first taste of journalism, commissioning him to write a book review. He was impressed by the young man's earnestness and idealism. 'He used to tell me: "You should be more ambitious, you can change this world." We used to joke that one day John Githongo would be president of Kenya.'

For Ali, John was remarkable for his attempt to get to grips intellectually with his stagnating country's predicament. The fall of the Berlin Wall in 1989 had ushered in a period of tumultuous change in Africa, in which a generation of post-independence autocrats faced increasingly strident calls for multi-party democracy. It was a drive Moi was doing his best to ignore, banning demonstrations and jailing opponents. 'Here was someone thinking his way out of a society that was tyrannically stable, a police state. There'd been a long period where the Asians, the *wazungu*, everyone had struck this compact with the political classes, comfortable accommodations had been made, everyone had his place in the system. In the 1990s people's heads changed. What drove John was the awareness that this society was on the crux of change, and the realisation that it is within your power to make things happen.'

Soon John was working as a columnist for *Executive*. His 'Political Diary' was published under a byline sketch that somehow managed to make a man in his mid-twenties look like a portly fifty-five-year-old. He was part of a small stable of irreverent young journalists with whom Ali experimented, pushing the boundaries of what had been until then fairly formulaic political coverage. 'He was someone who could say new things well, and provocatively. Sometimes he'd ramble a bit, but I'd cut him back.' John was among the first journalists to register the importance of the unfolding Goldenberg scandal, and also brought himself to the notice of the authorities by writing about the controversial decision to locate a bullet factory in Moi's home town. His phone was tapped, intelligence agents tailed him around town. But one of his biggest coups, in Ali's eyes, was to nudge the leaderships of Kenya, Uganda and Tanzania into agreeing to revive the East African Community, the region's defunct economic and customs union, by raising the issue during a series of interviews with the various presidents. It was a tribute to John's ability to draw his interviewees out. 'He could listen to you, and listen to you, and listen to you,' remembers Ali. 'He would quietly egg his interview subjects on until they ended up saying anything.'

Visiting Tanzania, John had come away impressed by former socialist president Julius Nyerere's success in forging a sense of nationhood amongst the country's disparate ethnic groups. Economically, Tanzania trailed Kenya, but its people seemed far more sure of their common identity. The opposite process seemed to be on display in Kenya, where, in the run-up to the 1992 multi-party elections, the opposition divided on tribal lines. What was worse, Kikuyu in the Rift Valley became the targets of a concerted campaign of ethnic cleansing. Not since the days of Mau Mau had the community felt so assailed. Kikuyu leaders gathered to raise money for the homeless, defence militias were pulled together, there were calls for the revival of GEMA. Driving through a tense Rift Valley, where the army roadblocks suggested a country under military occupation, John was shocked to be abruptly hushed when he addressed some women by the roadside in Gikuyu. 'Don't speak that language here!' warned a terrified hawker. 'We don't understand it here. Speak the national language!' Moi had always justified the one-party system on the grounds that Kenya was too potentially fractious, too ethnically diverse, to withstand the strains of multi-partyism. The clashes seemed designed to prove his thesis correct. 'Kenyans are now keenly aware of their ethnicity in a negative, destructive sense,' John wrote. 'We voted largely along tribal lines. We sowed the wind; I earnestly hope we shall not soon be reaping the whirlwind.'

Then, in April 1994, the genocide in nearby Rwanda brought Kenyans face to face with the full horror of what ethnic hostility, stoked by unscrupulous politicians, could do. While many Kenyan journalists, baffled by the bloodletting, preferred to ignore what was happening a mere hour's flight from Nairobi, John followed events closely. Could it happen in Kenya? This was not, he realised, a spontaneous flare-up rooted in long-running tribal hostility. It was an expression of top-level corruption, as political elites pushed competition over limited state resources to obscene lengths. 'Collective madness only happens by design,' he wrote in his column. 'Evil lies dormant, like a smouldering ember, in the human soul, and it can be fanned into flame by the most ordinary human passions – the passion for power, for wealth, for a good life for myself and my family.' If his education and upbringing hadn't already drummed in the point, the Rift Valley clashes and Rwanda's genocide highlighted the perils of a community withdrawing into an ethnic bunker.

When, in 1994, the Nation Media Group launched a business weekly called the *EastAfrican*, edited by Joseph Odindo, to be published simultaneously in Kenya, Uganda and Tanzania, many of *Executive*'s staff migrated to the new publication, including Ali. John had got there before him, adding a column in the *EastAfrican* to his activities, which now included hosting a political talk show on national television, running his own NGO – grandly baptised the African Strategic Research Institute – freelancing for the *Economist*, tinkering with a novel and editing a monthly publication for the Centre for Law and Research International.

Then a new opportunity presented itself, one that had a certain pleasing circularity to it. Back in the late 1980s, Joe Githongo had befriended Peter Eigen, the World Bank director for East Africa. With thirty years in development under his belt, Eigen had become exasperated at the way government sleaze kept sabotaging his organisation's work. Donor institutions, he felt, needed to get to grips with an issue regarded until then as taboo: corruption. He found a like mind in the shrewd Joe, who knew from personal experience how top-level patronage could tilt the odds against businesses that were out of political favour. 'He became, very soon, a hero for me, someone who was outside the system and challenging the establishment,' recalls Eigen. Joe became the African face of the initiative, which Eigen swiftly realised he would have to leave the World Bank to pursue. In 1993 Transparency International opened its main office in Berlin, with Eigen at its head, laying the foundation stone of a global anti-graft industry that has since blossomed into life.

TI chapters began opening around the world, but ironically its operation in Kenya, where the original idea was born, remained moribund. By 1999, however, with NGOs springing up across Nairobi and political debate opening up, the Kikuyu businessmen who sat on TI-Kenya's board felt the time had come to activate the branch under a new director. What more appropriate candidate could there be than the bright son of one of TI's founders? 'It was great fun. We had a budget and could do what we wanted, and corruption was a hot issue,' remembers John. Suddenly, local newspapers were full of stories about sleaze – from Kenya's abysmal world ranking in TI's yearly 'Corruption Perceptions Index' to a totting up of the breathtaking amounts top politicians contributed to *Harambee* fundraisers, cynical exercises in vote-buying and money-laundering.

John seemed unstoppable. But he might have been a little more cautious about taking the TI-Kenya directorship if he had overheard the conversation I had with a Westerner involved in TI's creation, who asked to remain anonymous. 'There was never any question in my mind,' he said, 'that John's father and his friends launched themselves so enthusiastically into anti-corruption work not because they believed in the cause itself, but because they had been economically boxed out by the Kalenjin.' There's a fundamental difference between supporting an abstract principle and backing a campaign that just happens, at a certain point in history, to mesh with one's personal interests. What happens when principle and interest no longer run parallel, but clash head-on?

A sociologist might look at the Githongo family and note a series of dislocations, physical and ideological, each serving to weaken the link between modern family and upcountry *shamba*. The Catholic/Protestant split, the Mau Mau/Home Guard schism, the first move from Kenya to Britain, the second to the melting pot that was Nairobi – each fractionally diluting the family's sense of ethnic belonging. Add to that the cosmopolitan nature of St Mary's, Joe's horror of tribalism, the importance of religion in the Githongo household, John's university years in Swansea learning the principles of Keynesianism and monetarism, his time spent in the subversive world of journalism, immersion in the NGO universe of global human rights, and the extent of the Mount Kenya Mafia's self-delusion in believing that they could control John Githongo becomes clear.

If being a 'good Kikuyu' meant putting his ethnic loyalties before all else, John was very, very far from that ideal. As he would later tell a reviewer, 'My employment contract did not say "Gikuyu Inc" at the top.' The remits of his compassion stretched far beyond what most of his elders and many of his contemporaries regarded as normal. He was a driven, highly moral, ethnically denatured young man who, if forced to choose – and 'Why should I?', one can almost hear him asking – would probably say that he thought of himself as an ethical and spiritual being first, a Kenyan second, a Kikuyu third.

Some would call this the opening of the mind that comes with a liberal education and an international upbringing, others would label it deracination, pure and simple. Forget too thoroughly, and you lose your anchor. 'Find out what language the Githongo family speaks at home,' a Kenyan political analyst told me, a touch sardonically, when I explained my book project. 'I'll bet you anything it's English.' In Kenya, language holds the key to identity, with the mastery of each language – first indigenous, then Kiswahili, then English – signalling a growing sophistication and an increasing distance from traditional ways. The answer to the question in the Githongo case, as it happens, is that family members speak a mixture of Gikuyu and English amongst themselves, adroitly manoeuvring their course through the various cultures that have shaped them.

Kenyan newspaper columnist Wycliffe Muga, a cheerful provocateur, sees John as an African so disconnected from his roots as to fall into the 'coconut' category – brown on the outside, white on the inside. 'The people John really wanted to impress were not the House of Mumbi, but the House of Windsor,' he chuckles. 'His loyalty to Western values – things like a belief in the importance of rules, transparency, honesty and accountability – was greater than his loyalty to the tribe. What the Mount Kenya Mafia didn't understand was that John wasn't a Kikuyu at all. He was a *mzungu*.'

To his recruiters, John represented a near-miraculous combination of skills and experience. But the forces shaping him were hardly calculated to produce the perfect presidential aide. They were calculated to produce the perfect whistleblower.

9

The Making of the Sheng Generation

'Whether Luo or Kikuyu, our children will not act the way we do.'

EVA GAITHA, director of a Nairobi coffin accessory company

Not only had John's talent-spotters misread their man. They had failed to register, in the complacent, careless way of the privileged, profound social and historical changes taking place around them, tendencies fuelling a national sense of exasperation with the old ways of doing things.

Langata Cemetery, which lies on the road linking the suburb of Langata with central Nairobi, is not the quietest of final resting places. Just across the busy road bordering its grounds is Wilson airport, the capital's second air terminal. It is a gathering place for traders sending bundles of the narcotic *khat*, grown on the chilly hill-sides around Mount Kenya, driven to Nairobi at breakneck speed to retain maximum freshness, and dispatched to twitchy customers in Somalia, Djibouti and Ethiopia. Returning *khat* flights drone constantly overhead, so close that graveside mourners get an intimate view of the aircraft's undercarriage and landing gear, fully extended prior to impact.

The cemetery is in constant, heavy use. With up to twenty-five funerals scheduled on an average day, three to four ceremonies are being staged simultaneously in different parts of the grounds at any one time. Styles and trappings vary, for there is no equality in death. Ceremonies in the 'permanent' side of the cemetery attract convoys of new cars, gleaming black hearses, and grieving relatives shelter from the sun under spotless white gazebos. Funerals in the 'temporary' section, where the soil is turned over and used afresh every fifty years, kick off with the arrival of a careering, inappropriately gaudy *matatu*, hired for the day by a fundraising committee, mourners crammed inside, coffin lashed onto the roof rack, with only the occasional tree offering shade. But both sides of the cemetery have one thing in common – an awareness of being on an industrial conveyor belt of death, whose brisk momentum is perhaps a little undignified but at least leaves mercifully little time to wallow in grief. 'It's very speedy, all over in about an hour,' says an undertaker. 'It's like a cocktail party. Everyone is standing, you're all uncomfortable, but it's over fast. That's why people like Langata.'

Demand is so heavy that the cemetery is running out of space. In the 'permanent' section, where members of the middle and upper classes end up below engraved marble stones, enclosed for all eternity in miniature villas with gravel lawns and iron gates, the grassy paths are being nibbled away by new burial plots. In the perfunctorily sign-posted 'temporary' area, used for the short of cash and for small children, new arrivals are being squeezed into spaces between existing graves, whose boundaries are discernible only to the cemetery officials and sweating diggers.

Once, burying a loved one in Nairobi was regarded as a near abomination by all but the colonials and Kenyan Asians, so rare that two French researchers described the city – inaccurately – as 'a capital without cemeteries' in a dissertation. Relatives paid for the dearly departed to be transported upcountry, back to the *shamba* they regarded as their true home. Only the destitute, nameless and friendless ended up in the capital's cemeteries. Now, demand so far outstrips supply that the price of plots in Langata has quadrupled in five years.

Undertakers shrug their shoulders and predict that the overcrowded cemetery cannot last much longer – 'In two years' time this place will be full,' mutters one – forcing the city authorities either to find a site for a new cemetery or to authorise the construction of private crematoria. And just as the former scarcity of local cemeteries once reflected the fact that the Kenyan nation was no more than an uneasy conglomeration of tribal statelets, the increasing tendency to bury the dead in the city marks a change in Kenyan society's sense of itself.

'The old folk always go upcountry,' says Benjamin Kibiku, director of Montezuma and Mona Lisa Funeral Services, the first indigenous funeral business to open in Kenya. 'With them, there is that feeling, "I have to be invested in the land of my ancestors."' Seeking a name that would be ethnically indeterminate, Kibiku baptised his company after a boyhood nickname and in tribute to his wife, 'because she's as beautiful as that portrait'.

Montezuma and Mona Lisa's motto – 'Service to the World' – is emblazoned on the wall of Kibiku's office, and he relishes the phrase 'one-stop shop' when describing what the company offers: coffin with satin trimmings, hearse, mourners' transport, gazebo, coffin-lowering apparatus – everything but the plot of land is included in the price. With quiet satisfaction, he shows off the forty-seat coaches, painted with the Montezuma logo, used to take city mourners back to their villages. 'I was the first person to come up with this prototype. It's specific to Kenya.' Just behind the back wheels, each bus boasts an idiosyncratic feature: an empty compartment with a neat glass porthole, for stowing and easy viewing of the coffin during the bumpy, hazardous trip home. No one wants a coffin falling out in transit.

Montezuma's coaches are permanently booked, Kibiku says, yet he's noticed the beginnings of a generational divide. 'Young families feel that they met in Nairobi, they married in Nairobi, they have no interest in upcountry. So when members die, they are buried in Nairobi.'

The change is at its most obvious amongst the adaptable Kikuyu, least evident in residents from tradition-bound Western Province. Population pressure in Central Province means families are anxious not to waste farming land. A body on the premises not only gets in the way of planting, it makes land difficult to lease or sell. There's also a question of cost, with an upcountry funeral setting a Nairobi-based family back up to 180,000 shillings, more than many can afford. But underlying all those considerations, Kibiku acknowledges, rests his customers' subtly shifting concept of what counts as home. 'An attitude change is under way.'

That shift seems inevitable when one looks at the figures. Like other African nations, Kenya is experiencing vertiginous urbanisation, as shifting climate patterns, the subdivision of plots, soil degradation and mechanised farming push those who will never inherit land and are no longer needed to work it towards the city. In 1962, one in twelve Kenyans lived in urban centres; by 1999 the figure was one in three, with half the population expected to be city-based by 2015. 'Whenever I go upcountry I'm always amazed at how empty it is,' says a Kenyan journalist born in Luoland. 'Only drunks, idiots and the old, only failures, stay behind in the rural areas.'

Originally designed for just 200,000 inhabitants, Nairobi now holds 4–4.5 million. It has more than doubled in size in the past five years, giving it one of the highest growth rates of any African city. That growth consists almost entirely of the poor, whose shacks have filled what were the green spaces in a network of loosely connected satellite settlements. Nearly two out of three of the capital's inhabitants occupy the two hundred resulting slums, a steady source of income for City Council officials, too busy levelling fantasy 'taxes' on the unauthorised dwellings to want to alter the status quo. Among the most squalid the continent has to offer, these settlements nuzzle against well-heeled residential areas in provocative intimacy. 'What's striking about Nairobi is that each wealthy neighbourhood lies cheek by jowl with a slum,' remarks former MP Paul Muite. 'It's almost like a twinning arrangement. Poverty and wealth stare each other in the face. And that's simply untenable. Those slum-dwellers know what they're missing, they're educated now. I tell my wife: "There's no way, long term, those guys are going to accept to die of hunger when the smell of your chapattis is wafting over the wall."'

The biggest slum is Kibera, virtually an obligatory stop these days on visiting VIPs' itineraries. Kibera, bizarrely, lies within a tee shot's distance of Nairobi's golf club. Aerial photographs show the neat green medallion that is the club abutting what looks like a brown sea of broken matchsticks, in fact the corrugated-iron *mabati* roofs of between 800,000 and 1.2 million residents, prompting the immediate mental query: 'Why don't they just invade?' Kibera is where the phrase 'flying toilets' was added to the English language, a description of the method used to dispose of faeces – dump it in a plastic bag and throw it out of the window – by residents who couldn't be bothered walking to the public latrine. Yet while the slum does not boast regular electricity, tarred roads or clean water, it offers hope of a different kind. If your children miraculously survive to the age of five in Kibera, they will go on to receive a far better education than their rural equivalents, and in that education lie untold possibilities.

By the late 1990s, many analysts were confidently predicting that population trends alone would accomplish what Kenya's presidents had failed to achieve with their national anthems, independence days and flag salutes: a true sense of nationhood. Nairobi's first slums were mono-ethnic, the result of colonial attempts to corral Africans into distinct, controllable areas during the Emergency years. The newer ones started out that way, but the phenomenon didn't last long. Often dubbed a Luo settlement, Kibera itself actually contains forty-two separate tribes, 'all doing their jig together', as an official from the UN's Habitat told me. Ethnicity blurred in playgrounds, schools, universities and offices. 'When people first arrive in Kibera, they initially go to where their people are and look for work. They arrive with nothing, so to cut costs they sleep six to a room. The longer they live together, the more they fuse. They are forced to share meals, they share *mandazi* [doughnuts]. They mix at school, at political rallies, at prayer. The old people are the problem. But the kids don't know whether they are Kikuyu, Luo or Kalenjin, whether they are from Tanzania, Uganda or Kenya.'

Even in the space of the dozen years I reported on Kenya, it was possible to log a fundamental shift in the way Nairobi's residents viewed themselves. When I first arrived, the easiest way to discover someone's ethnicity was to ask where they came from. Nobody ever said 'Nairobi'. Even those born and brought up in the capital felt they were essentially *from* somewhere else. Historically a mere junction between Kikuyuland, Maasailand and Ukambani, Nairobi just happened to be the place you received an education, held down a job or brought up a family. It remained a form of no man's land, an accidental city, commanding little pride, strictly temporary. By the end of the Moi era, the reaction to that same question was different, and as often as not came with a defiant stiffening of the spine. 'Where am I from? I'm from Nairobi,' a student would say, or: 'Look, I consider myself a Kenyan.'

Kenya's demography makes radical change inevitable. A staggering 70 per cent of the population is below the age of thirty. That statistic, shared with many African nations, is as hopeful as it is terrifying. And the fact that those youngsters do not think in the same way as their parents is highlighted by the fact that they no longer speak the same language. English and Kiswahili might be Kenya's official languages, but pupils tumbling out of school and students in the university canteens chatter to each other in Sheng, to their teachers' despair. A witty, cheeky, freewheeling *Clockwork Orange*-style brew of Kiswahili, English and indigenous Kenyan languages, with added dollops of reggae jargon, American slang, French and Spanish, Sheng originated in Nairobi's Eastlands slums in the 1980s. Adopted by *matatu* touts and rap artists, it radiated along the taxi and bus routes, spilling over into Tanzania and Uganda, moving from one urban centre to another. So popular has it become that sending an email or text in Kiswahili or English rather than Sheng is considered disastrously uncool by anyone below the age of twenty. Infiltrating radio stations, it has forced its way into national newspapers and spread its tentacles across the internet. Kenyan publishers promise future books in Sheng, it features large in advertising slogans – why, it even crept into Kibaki's speeches.

This rogue language's popularity is something of a contradiction: Sheng was originally invented to exclude the puritanical parents and ball-breaking teachers who threatened to prevent a younger generation having a good time. Kenyan youths wanted to be able to discuss their sexual adventures, hangovers and boozy nights in their elders' presence without the latter cottoning on. A language in a hurry, it did away with the grammar and spellings slowing Kiswahili and English down, and had the same ingredients of topical humour and impish wordplay as France's Verlan or London's Cockney rhyming slang. Breasts are '*dashboad*' – from the English 'dashboard'; protruding buttocks are baptised '*to be continued*', a Casanova is a '*lovito*'. One of the many words for party is '*hepi*' ('happy'); a cigarette is a '*fegi*' (from the English 'fag'); a friend a '*beste*', and the term '*jigijigi*' – sex – needs little explanation.

Constantly inventing new terms was part of the game, allowing the speaker to show off his ability to dip into five or six languages without pausing for breath. As a result, a web-based Sheng–English dictionary, still being compiled, gives at least thirteen alternatives for 'girl' – including '*chic*' (Spanish/American), '*chipipi*' (Luhya) and '*mdem*' (French) – seven for 'money', and five each for 'house' and 'school'. Sheng spoken in Nairobi's Dandora slum differs from that spoken in the city's Eastlands area, and because the language is always on the move, shifting like a Chinese whisper from mouth to mouth, it dates fast. On their return home, Kenyans in the diaspora find the Sheng used in their blogs no longer matches the Sheng spoken by childhood friends. Incomprehensible not only to parents but even more so to staid grandparents back in '*shags*', 'deep Sheng' is a barrier behind which the new generation can hide its secrets.

Traditionalists shake their heads, seeing the threat of dissolution. 'Let Sheng be left to matatu touts, drug pushers, hopeless hip-hop musicians and school dropouts,' argued a columnist in the *Standard*, slamming it as 'linguistic jingoism'. But the dialect probably represents exactly the opposite, a force for national unity. Supporters point out that whereas Kiswahili and English were brought to Kenya by Arab traders and English settlers, Sheng was an indigenous invention. 'At last, here is something truly ours for once, which unites us, and which we haven't inherited,' wrote a defender. As a language of the poor embraced by children of the elite, as anxious to sound trendy on the playing fields of their private schools as any slum urchin, it is a class leveller. Writer Binyavanga Wainana sees Sheng as the expression of a youth revolution which militates against the sharp tug at the ethnic heart-strings many Nairobi residents experience with the onset of maturity. 'As you get older, entering into marriage and having children seems to tribalise you. All those ceremonies, marriage arrangements, land issues; those decisions about which language to bring your children up in and which school to attend; they activate something in people they didn't know they had.

'Sheng has given us all a safe language to speak. There's a kind of hopefulness to it, a feeling of establishing a sensibility which encompasses tribe, is working-class, inward-looking, philosophical.' Perhaps, on a continent in which identity and language are so interlinked, in which almost every African seems to have mastered four or five languages; and with each language, four or five different ways of interacting with others, only Sheng, with its rich, shifting mix of associations, can express the kaleidoscopic entity that is the modern Kenyan.

If the Sheng generation is more streetwise, technologically savvy and sexually knowing than its elders, it also has a radically different awareness of its rightful place in the world. As the Kenyan middle class expanded, so did the numbers of youngsters sent abroad to complete their training. Parents dispatched their offspring hoping they would learn how the world worked and win the keys to Western-style prosperity. But those who return – and a disconcertingly high proportion choose not to – look at their continent and their kith and kin with the pitiless, unforgiving eyes of the youthful idealist. They have done the maths, they understand economics and have read the newspapers. They are all too aware of how much better things work elsewhere, painfully conscious of the extent to which, in foreigners' minds, Africa is logged in the 'basket case' category. And for that they blame the very people who paid for their eye-opening educations.

* * *

Conrad Marc Akunga, whom I met in early 2006, was one example of the young iconoclasts springing up in modern Kenyan society. Tall and skinny, with the awkwardness of an overgrown swot, he looks exactly what he is: a computer geek with an instinctive empathy for the world of gigabytes and downloads. A blogger on Kenyan affairs, he met up with Ory Okolloh, a female graduate of Harvard Law School, in the wake of the 2002 elections, as disillusionment set in. Together they decided to set up www.mzalendo.com, a website aiming to make Kenya's parliament more answerable to voters. One of the incoming MPs' first acts was to hike their own salaries, making them among the best-paid parliamentarians in the world, let alone Africa. Furious civil society groups pointed out that while the lawmakers benefited from monthly earnings 270 times the average, Kenya's parliament, in terms of days attended and bills passed, was one of the least productive on the planet. 'Ory and I used to get together and rant: it was "these guys, these guys, these guys",' says Marc. 'It got to the point where over breakfast one day we agreed to take action. The first thing we needed was information about who "these guys" were.' Out of the desire to do something other than whinge, the idea for Mzalendo – 'patriot' in Kiswahili – was born.

Originally modelled on the Westminster system, Kenya's 222-seat parliament is in theory transparent to the public. One form this openness is supposed to take is the Kenyan version of Hansard, the written transcript of parliamentary proceedings. In fact, parliamentary officials treat access to Hansard as a privilege rather than a right, and the paper transcript is, of course, of little use to rural voters wanting to know what their MP gets up to in the capital. 'These guys talk loud in public, but what they do inside that chamber isn't known,' says Akunga. 'Some have never once spoken in parliament. It's your right, as a voter, to know that.' The website gave, when it could, profiles of MPs, a rundown of their educational qualifications (often a sensitive subject), details of which committees they sat on and which motions they proposed. It provided telephone numbers and postal addresses, allowing dissatisfied constituents to pester their elected representatives in person.

Its two founders launched the project with only token assistance from donors. With no staff and no premises, Mzalendo didn't actually need money, Akunga told me, demanding instead time and commitment. Akunga, whose day job was with a Nairobi software company, provided the technical knowhow. Ory, who had moved to South Africa, focused on content, cajoling parliamentary officials into providing back copies of Hansard.

Mzalendo, it has to be said, will never be a YouTube favourite – it is far too worthy to make for gripping reading. But in the duo's eyes, the four-to-six-hundred daily hits the website gets justify its existence. The two hope to counteract what they see as a national tendency to tut-tut briefly over human folly, give a resigned shrug, and move on. It is a characteristic that gave exploiters an easy ride, allowing a small group of players to circulate like soiled clothes in a washing machine. Jumping from party to party, campaigning against policies they championed until recently, politicians rely on general amnesia to survive one scandal after another. 'Kenyans tend to forget easily and forgive easily; it takes just a few weeks. We hope the website will work against that.'

Most MPs have done their best to ignore the website. That indifference reinforced Akunga's cynicism towards lawmakers, who he blames for a steady poisoning of the political climate. 'If Kenya is ethnically polarised today, it is these guys who are at the root of it. You grow up in Nairobi and you play with everyone and then at university you suddenly start hearing people say, "They're out to get us." If the MPs just shut up, we'd sort it out, but instead they keep fanning the flames. Even educated fellows, professors, say the most unsavoury things quite openly. They seem to forget that microphones have memories.'

The project was an example of how one expectation feeds another, furtive hopes mutating into strident demands as the citizen's sense of what is his due expands. Akunga got his first heady taste of political activism in the 2002 elections, when the boss of his software company designed a programme to collate the results, a plan hatched to prevent vote-rigging. Party officials rang in from the constituencies with the tallies, and staff immediately typed them into the computer. 'We worked all through Christmas, working till 4 o'clock in the morning, working so hard we didn't even have time to go out and vote ourselves,' recalls Akunga. 'I remember when the votes came in and we saw the final result, I had this amazing feeling: that we had played our part in bringing that about, we had done our bit. There was this incredible sense of euphoria.' That intoxicating experience had carried all the emotional force of a religious conversion. Mzalendo.com was Akunga's attempt to keep the novel sensation of being part of something bigger and better than himself alive.

Akunga's irreverence was magnified twenty times in another Kenyan who seemed to represent what was to come. I'd first heard of Caroline Mutoko at a lunch in Muthaiga. A guest was complaining about a verbal lashing a colleague had been subjected to on Kiss FM, one of the capital's popular radio stations, by what sounded like a razor-tongued virago. A government minister joined in: he too had borne the brunt of the harridan's ire. He enjoyed a reputation as something of a progressive, so I was surprised to hear him casually mention that he had tried, without success, to persuade Kiss's management to take the presenter off the air. He shook his head. 'It's beyond a joke. She simply goes too far. The woman has to be stopped.' This Bitch from Hell, I thought, was definitely someone I wanted to meet.

I told Mutoko the story when I met her, and she gave a mirthless laugh. 'There's not a politician who likes me. Not one.' She shrugs. 'And I don't mind. They are all extremely charming when they meet me in person, but I know that behind my back they're saying: "Oh my God, get her oesophagus."'

Manicured, carefully coiffed and sporting the very latest thing in sunglasses – a 'parasite' model which clung to the face rather than hooking round the ears – Mutoko carried with her the near-visible aura of celebrity. She is not a big woman, but gives off an air of ineffable self-confidence, much of which can be traced to the timbre of her voice. Many Nairobi broadcasters speak a very Kenyan form of English, with the stress placed on syllables no Briton emphasises. Not

Caroline. Her English has the crisp precision of a Kenyan Joanna Lumley, a quality she attributes to the Irish nuns at her school who made their pupils read long passages aloud. As warm as chocolate, low, smooth, self-assured, hers is a voice perfect for radio, letting her listeners know they are among friends.

It is such a purr that the violence of the sentiments it expresses are doubly shocking. Mutoko talked, over a lit cigarette, about wanting to slap politicians in the face, of being 'pissed off' by the powers that be and of 'butchering' those who dared repeat 'the same old crap' on her programme. Transposed to the airwaves, the approach, in a country hamstrung by etiquette, has won her the status of one of America's 'shock jocks'. Like them, she sometimes appals even her most fervent fans. Like them, she is simply too entertaining to miss, and her *Big Breakfast Show* is one of Kenya's most popular.

She'd migrated to Kiss FM after becoming exasperated by the triviality of her job at rival Capital FM, where she was 'an expert on Robbie Williams, and there was *nothing* I didn't know about the Spice Girls'. Neither Capital nor Kiss would exist had it not been for Moi's reluctant liberalisation of the airwaves in the mid-1990s, a move which marked the waning of the deference my piqued fellow guests in Muthaiga felt was their due. 'We live in a country where people in power don't realise they are actually public servants,' says Mutoko. 'When you're a politician in Kenya you're used to grovel, grovel, "honourable", "honourable". You expect to be treated like a demigod, so it's very hard when people say, "Screw you." This is a scary time for politicians.'

Talking to Mutoko, one sensed a roiling, restless fury, a huge impatience finding expression after years of control. She sees herself as mouthpiece for an entire nation whose patience has snapped. 'This country is on such an amazing high it can't be stopped. Kenya is awakening. I can hear it in the phone calls we get. People ring and say, "This road has been worked on since September, it's now March" . . . They call in to bitch about not having water for three days, not having power. That never used to happen before. We've become a whole lot more questioning. You can't sell me shit.'

Like Mzalendo's founders, much of Mutoko's bolshiness lies in her awareness that Kenyans, through their passivity, have contributed to their downfall. 'Half our problem in the media was that we self-censored. You self-censor and then you wake up one day and realise the way things are is your fault.'

Convinced that an ossified political class was trailing far behind its public, Mutoko, when I interviewed her in April 2006, was encouraging young people with no previous experience to stand in the next elections. Prospective candidates, including youngsters from Nairobi's slums, were invited onto Kiss FM to explain their manifestos. It was a high-risk strategy: 'But I would gamble anything on difference. I already know what your track record is,' she said, rhetorically addressing a member of the old guard, 'and it's crap. Your track record is garbage.'

A Kamba by birth, Mutoko should in theory have been rooting for Kalonzo Musyoka, former foreign minister, presidential aspirant, and a fellow tribesman. 'People stand next to me in bars and whisper: "If Musyoka gets in, you know as a Kamba you could get a good position, because it's our time."' In fact, she scorned an approach which would have made a nonsense of the meritocratic principles on which she had based her career. 'The whole "our time to eat" line is the worst thing that ever happened to Kenya. You'd like to find the first person who ever used it and drive a stake through their heart.'

It was impossible to separate Mutoko's political stroppiness from what was, essentially, a feminist itinerary, one that appeared to have largely despaired of the African male. Single and childless, she was immensely proud of the fact that she lived in a house paid for by her salary and boasted a share portfolio built from her earnings. 'I'm a Nairobi woman who has finally found my feet and my voice. I'm not looking for anyone to complete me.'

Leaving Mutoko that day, a sudden image came to mind: of a tightrope walker who has never experienced a serious fall, stepping forward without a net. Chin up, back straight, the acrobat gazed into the middle distance, never looking down. 'The day I give in to the fear, I might as well resign,' she had told me when I asked about a court case a minister had brought against the station. The velvet-toned presenter, I suddenly realised, was one of the few people I'd met who simply didn't seem to know the meaning of the word.

By demanding that Kenya's multi-party democracy should possess substance as well as form, Mutoko, Akunga, Okolloh and their ilk were taking on an entire school of political thought about Africa. Their convictions challenged those cynics who dismissed John Githongo's anti-corruption efforts as the naïve projection of inappropriate '*mzungu*' values onto an African nation where they were doomed to fail. If John was a 'coconut', he certainly wasn't the only coconut in Kenya.

Could John, then, be relied upon by his colleagues to keep quiet, given what he represented? At first glance the notion of a challenge to the establishment emerging from within the upper class, the very social group that benefited most from the status quo, might seem counterintuitive. But the aristocratic scion who chooses to live as a pauper, the class rebel whose antagonism towards his peers is based on the most intimate of understandings, is a well-established historical phenomenon. Discussing why someone of John's lofty caste might choose a 'deviant' path, Dr Tom Wolf, a US analyst living in Kenya, cites the examples of Lenin and Fidel Castro, 'both from well-established, upper-middle-class families . . . who nevertheless re-engineered themselves into the most ferocious of revolutionaries'. Mahatma Gandhi came from a long line of statesmen, Che Guevara was of aristocratic descent, John and Robert Kennedy were born into a family of immense wealth, much of it shadily acquired. Growing up close to power, Wolf argues, probably ensured that John was 'less in awe of those wielding it' than a Kenyan contemporary from a more humble background, anxious to assimilate. 'In that sense, his "class" heritage encouraged independence of thought and action, rather than sycophantic loyalty.' 'Less in awe' is putting it mildly. 'Generational contempt' might be a more accurate term. 'My parents' generation are the reason we are in a mess,' Mugo Githongo, John's brother, once told me. 'We have nothing to learn from them.'

There was also a specifically Kikuyu shape to John's revolt, and it took outright oedipal form. With the oil crunch of the 1970s and 1980s, which coincided with Moi's abandonment of any attempt to win the acceptance of the business elite, many Kikuyu heads of families lost their jobs. Patriarchs who had grown fat on the cream of the Kenyatta regime, bragging about their deals in the bar, suddenly found themselves out of work, disrespected at home and with time on their hands. 'In the Kikuyu community you are raised hearing ponderous voices saying "the family is this, the family is that". But as you grow there's a straining of the social contract, it begins to crack,' says Martin Kimani. 'You become aware of this tension in the house, this anger between your mother and father.' Just as Kikuyu women took the helm in the absence of their menfolk during the Mau Mau Emergency, these church-going mothers came to the family's rescue. Working as nurses, running milk-trading schemes and sewing circles, they counted the pennies as their husbands wasted pounds on beer, mistresses – who gave the men the respect they no longer received at home – and deals that failed to miraculously revive the family fortunes.

Kikuyu sons, who traditionally enjoy a special bond with their mothers, absorbed their anger at wastrel husbands, the humiliation as rumours of girlfriends and illegitimate children drifted back to the family home. Like doting gangster bosses who send their daughters to convent schools, only to be horrified when the girls turn judgemental on them, many Kikuyu fathers endured the final irony of seeing their offspring turn against them, rejecting their entire code of behaviour. 'The generation that sent its sons to schools where they got a good Catholic education were hoist by their own petard. The NGO values, the desire to build a better, more virtuous Kenya, were all sown at home,' says Kimani. 'It's a very complex family drama.' The children in these families often only grasped the full extent of a paterfamilias's perfidy at his funeral, when they would discover, standing at the graveside, a previously unknown second family, complete with second widow, grieving children and claims to the estate.

If these were some of the broad trends that meant John's allegiance could not be taken for granted, there were also some simple pragmatic reasons why a man of his background might choose to act with unprecedented recklessness.

Under both Kenyatta and Moi, the government's hold on the Kenyan economy – whether in terms of civil service jobs, parastatal posts or contracts up for grabs – was so vast that alienating the president virtually meant financial ruin, as Joe Githongo and many Kikuyu entrepreneurs knew only too well. You could be firmly entrenched in the private sector and *still* be crushed by the hostility of the powers that be. Centralised systems of power are like onions: each layer faithfully mimics the core. Make an enemy in the top echelons of government, and business mysteriously dried up, public tenders no longer went your way, your goods took an age to clear customs, your phones were never connected and the tax inspectors took a sudden, obsessive interest in your accounts. There was simply no getting away from State House. Fall foul of the president or one of his cronies and the only thing to do was to hide yourself away and pray for either a change of regime, however long that took, or eventual forgiveness.

Scoured by the brisk winds of economic liberalisation in the 1980s, the Kenyan state shrivelled. Once a bright graduate would have automatically regarded a job in a government ministry, with its grading structure, pension scheme and subsidised housing, as the summit of his ambitions. Although structural adjustment reforms were often met with horrified cries from the likes of Oxfam and Christian Aid, the expansion of the private sector, the birth of a vibrant civil society and the blossoming of the media opened up a range of interesting new job opportunities. Who cared about keeping on the right side of State House when it was possible to join a South African research institute, work for a Nairobi-based multinational, or set up an NGO to garner well-paid consultancy work for foreign donors? John's experience at Transparency International had taught him there was life outside the state sector. Joining government had been a choice, not a necessity. What was more, the lucky fact of his birth in the United Kingdom meant that he enjoyed automatic residency there, and could work anywhere in Europe. That prospect seemed intriguing rather than appalling to a man who felt himself to be a citizen of the world. In the shape of John Githongo, the Mount Kenya Mafia was dealing with the first generation of Kenyans whose members were financially free to follow the dictates of their conscience.

And then there were the individual peculiarities of the man himself. Ever since his childhood, John had come under huge pressure from those he loved: to excel academically, to join the family firm, to get married to a nice Kenyan girl, have kids, get involved in politics. He had ducked and swerved with skill, charting a highly individual path through the expectations of others. As a result he found himself that strangest of beings: a forty-year-old African bachelor with neither wife, household nor children. By the conventional standards of macho Kenyan society, John Githongo was weird, little short of a freak. In pre-colonial Kikuyu society, where only those regarded as having achieved something of lasting worth in their lives were honoured with burial, the bodies of unmarried men were carried outside the homestead and left for hyenas to devour. Dying incomplete, they might never have existed. Yet John was willing to defy those norms rather than neglect what really made his heart beat faster: the life of the mind.

Worryingly for the presidential coterie, John's eccentricities were all of the sort to close down possible avenues for applying leverage. For most Africans, the weight of the extended family serves as ballast, tethering their ambitions firmly to the ground. When confronting our destinies, we each tot up what we stand to lose and what we can bear to surrender. John had no direct dependants. Should he choose to commit career suicide, no child's education would be at risk, no ambitious wife's laments echo around the family home, no in-laws make impassioned appeals to his common sense. Gloriously self-sufficient, John Githongo was free to take whatever decisions he chose.

Even if the Mount Kenya Mafia's members were not given to probing motives and analysing underlying social shifts, an episode in John's past really should have caused them some concern.

In 1998, before John moved to TI, he had briefly agreed to launch a regional political magazine on behalf of a Kenyan NGO called SAREAT (Series for Alternative Research in East Africa Trust). Headed by political scientist Mutahi Ngunyi, SAREAT won generous funding from the Nairobi office of the US-based Ford Foundation. But the project, John came to believe, was being used to front a scam. He suspected Ford's local programme officer, a Zimbabwean called Jonathan Moyo who would go on to serve as Robert Mugabe's information minister, of plotting with Ngunyi to misappropriate tens of thousands of dollars of Ford funding. Carefully documenting what had happened, John walked away from the project. Initially, recalls former work colleague Milena Hileman, he had no plans to report the issue to Ford's senior management. 'He was very upset, but he simply didn't trust the *wazungus* to do anything about it. I kept nagging him, saying, "Not all *wazungus* are the same. These are good people and they need to know."' In the end, the two flew to New York to brief Ford's vice president, and the foundation sued both Ngunyi and Moyo.

A practice run, in many ways, for Anglo Leasing, the SAREAT episode was not exactly an encouraging sign for anyone hoping the new permanent secretary would show leniency towards those caught with their hand in the till.

10

Everything Depends on the Boss

DAVID FROST: So what in a sense you're saying is that there are certain situations . . . where the president can decide that it's in the best interests of the nation or something, and do something illegal?

RICHARD NIXON: Well, when the president does it, that means that it is not illegal.

Excerpt from interview, aired 19 May 1977

As the long rains spluttered to an end in June 2004, Kenya's anti-corruption chief was aware of a range of possible scams being hatched across various government departments. Yet he decided to train his focus exclusively on Anglo Leasing, for unashamedly political reasons. There were only so many projects his office could handle. Had he channelled his energies into cleaning up a Kamba-led ministry or Luhya-headed department, it would have looked like ethnic targeting of the most blatant kind. 'I had to make choices. I knew of at least one other minister who was a complete crook, but I was a Kikuyu, in a Kikuyu government, and people around me were saying, "It's our turn to eat." You have to start at home.' Although he knew few fellow kinsmen would see it as such, his approach was a form of protective ethnic loyalty – his interpretation of what it meant to be a 'good Kikuyu'. In the tribe's long-term interests, a fundamental principle needed to be established. Ever since Kibaki's inauguration, cynics had been predicting a return to the abuses of the Kenyatta era, when the Kikuyu had enriched themselves without compunction. The new government had to prove such predictions false, show that such practices had become obsolete in the new Kenya, if history was not to repeat itself. 'Moi's term in office wouldn't have been possible without widespread resentment towards the Kikuyu. Moi was a Kikuyu blunder. The richest and biggest tribe has to be the most magnanimous.'

John's suspicions of key colleagues did not necessarily mean his entire venture was holed below the waterline, he told himself, reciting the Chinese proverb 'A fish rots from the head down.' As long as the president was still committed to the fight against sleaze – and that was certainly what John told any diplomat, journalist or NGO director willing to listen – his ministers could be tackled and thwarted. The simple matter of John's geographical location inside the white-colonnaded edifice of State House held the key, he had come to realise, for it guaranteed access to this least ideologically steadfast and most indolent of presidents. 'On several occasions, John told me, "The battle will be where my office is,"' remembers Richard Leakey. 'John said, "If I can walk down the corridor into the president's office, then I can continue. If I have to move downtown, it's over."'

As family friends and prominent politicians virtually queued up to urge him to stop digging, John was also coming under pressure from the opposite direction. Alarmed senior officials were nagging him to now turn his attention to the second suspect Anglo Leasing deal, which involved the building of a forensic laboratory. His more lowly informants were not letting up, either. The flood of leaked information threatened to drown him. At times he felt as if his overenthusiastic agents were running him, rather than the other way round. 'I became overwhelmed. I felt I could not satisfy these individuals.'

The pressure had an effect. On 4 June, after only ten days of playing the 'softly-softly' game, John violated his self-imposed ceasefire. He wrote to Andrew Mullei, governor of the Central Bank, asking him to stop all further payments to Anglo Leasing and requesting details of money transfers to the firm to date. The letter was a measure of his distrust of the permanent secretaries, as he had already written – in vain – to Joseph Magari at the finance ministry making the same request. As John was bracing himself for a response to this salvo, he was leaked a letter that put his efforts into chilling perspective. What he had glimpsed so far, the document made clear, was the merest tip of the iceberg. The Anglo Leasing affair was far more than a couple of dodgy contracts with a shadowy Liverpudlian company. It was a modus operandi taking place beyond parliamentary scrutiny, across two administrations, enthusiastically replicated by its inventors in confident expectation of huge profits.

The letter, drafted by the Central Bank governor, was addressed to the finance ministry and asked for confirmation that Mwiraria had authorised payment on a long list of contracts. The list embraced the two projects John already knew about, but included sixteen other contracts – eighteen in all – which had variously been signed off by the finance, transport and internal security ministries. Kenya's auditor general would later draw up a similar, even more detailed inventory. All the contracts could be described as 'sensitive'. Military- or security-related in nature, they included a digital multi-channel communications network for the prison service, new helicopters, a secure communications system for the police, that state-of-the-art frigate for the navy, a data network and internet service satellite link for the Kenya Post Office, a top-secret military surveillance system dubbed 'Project Nexus', an early-warning radar system for the meteorological department, and so on. Twelve of the contracts had originally been signed under the previous Moi regime and carried over under NARC. The other six had been signed by the new government. The list gave dates and values. Added together, the auditor general would calculate, the eighteen contracts were worth a gulp-inducing 56.3 billion shillings ($751 million).

One of the problems that would confront those trying to whip up public ire over Anglo Leasing – the same hurdle faced by those trying to stop Goldenberg in the 1990s – was that for a Kenyan public accustomed to calculating daily expenses in hundreds and thousands of shillings, the billions involved had an abstract, intangible quality. Ironically, the bigger the figures, the less relevance they seemed to have to ordinary lives, making it difficult for civil society activists and opposition MPs to persuade voters that Anglo Leasing deserved their attention, taking a vast bite, as it did, out of their taxes and hiking their daily costs. Like a tourist attempting to capture the vast cathedral of St Paul's in his viewfinder, the ordinary Kenyan struggled to focus on the gigantic object looming unexpectedly before him.

In fact, the value of the eighteen contracts amounted to 5 per cent of Kenya's gross domestic product, and over 16 per cent of the government's gross expenditure in 2003–04, the period in which the six NARC-era contracts were signed. It easily outstripped the country's total aid that year ($521 million), and represented three quarters of the amount the hard-pressed Kenyan diaspora annually sent back home. The campaigning anti-graft organisation Mars Kenya would later calculate that the funds involved were the equivalent of 68 per cent of what the finance ministry allocated to infrastructure in 2006, and thirty-seven times more than it allocated to water projects in Kenya's arid lands. The American ambassador came up with an even more depressing figure: the money would have been enough to supply every HIV-positive Kenyan with anti-retrovirals for the next ten years.

While not on Goldenberg's gargantuan scale, Anglo Leasing was still big enough for its knock-on effects to impact on every Kenyan, no matter their social status. The scale of the thing made the nervousness John had encountered from suspected key players suddenly comprehensible. There was plenty to be uneasy about.

When I was stationed in Kenya as a foreign correspondent, I worked out of the international press centre in Chester House. The building is located in Nairobi's business district, next to one of the city's least salubrious nightclubs, on a scruffy street that is lined at night with prostitutes in tight white miniskirts. The press centre was often a very quiet place, because the Western reporters spent half their lives on the road, relying on Kenyan staff to pay their bills, update their files and order stationery during long absences in Rwanda and Burundi, Ethiopia and Somalia. As the years passed, many of us began to suspect that our office managers, routinely trusted with huge volumes of cash, were getting a little sloppy. Going through my receipts, I'd be taken aback to see the price we'd paid for a few reams of photocopying paper, a plastic binder, or some ink. I'd pause, stare at the receipt, dark suspicions stirring. But there was always something more pressing that needed doing. If I couldn't rely on my office manager to do his job, my entire operation became unsustainable. Then, one day, I decided to check a suspicious receipt against the high street price. We'd paid three times as much. Furious, I confronted my office manager, planning to storm to the shop in question to have it out with the owner.

'Where the hell did we buy this? Whoever sold it is a complete crook.'

My plan for a showdown immediately stalled. Looking uncharacteristically flustered, my office manager explained that, like his colleagues in the building, he routinely bought stationery from a supplier who went door to door in Chester House, rather than venturing into Nairobi's business district.

'But you've been paying him three times the going rate!'

'I didn't realise. I thought he was giving us special prices.'

'They were *special* all right.'

'I'm sorry. This will never happen again.'

For a moment, I almost fell for it. Who was this nameless, faceless, nebulous salesman who had been conning Chester House staff? I tried to remember if, during my stints in Nairobi, any office supplier had ever come knocking on my door. I could remember the Sudanese rebel spokesman with his acronym-spattered declarations, the charming amputee who begged for alms, the student hungry for freelance work. But no shadowy stationery supplier. I quizzed colleagues down the hall. None of them remembered him, either.

Of course, there was no such man. He was a ghost, conjured into existence in the mind of my canny office manager, whom I eventually sacked when the fiddling became too obvious to ignore. Capitalising on my inattention, he had got together with a small army of obliging high street shop staff to fabricate inflated receipts. The shop owner, likely to be a Kenyan Asian, would get the official price. The difference between that and what I actually paid would be split between my office manager and the shop assistant, probably a kinsman – two sons of the soil ganging up satisfyingly to con the *mzungu* and the *mhindi* in one fell swoop. There was nothing original or complicated about the scam: it was the oldest trick in the book. It was, as it happened, almost exactly the technique used in Anglo Leasing.

While Goldenberg involved financial stratagems so complex even economists had difficulty grasping them all, Anglo Leasing was so simple a child could master the technique. It was a classic procurement scam, needing only two parties, although for it to work one of those parties had to be at the top of government, powerful enough to silence doubting minions and ignore institutional checks and balances. Another vital component was the military and security nature of the contracts. In every other sector, contracts had to be put to open tender, forcing even the greediest of suppliers to stay within the realms of the reasonable. Advertisements were placed in newspapers and trade journals, competitive bids collected, technical competence established, and the companies who came forward knew they might be subjected to due diligence before the relevant ministry made its decision. Any irregularities risked being picked up by opposition-dominated parliamentary committees and scoop-hungry journalists. The only area escaping such scrutiny was the security and intelligence sector, where single-sourcing, opaque negotiations and loosely worded contracts could all be justified on the grounds of 'national security'. Could the government really be expected to tout for a counter-terrorism centre on the open market? Or to publish specifications for a new prison communications network in the press? Of course not. It would put Kenya itself at risk.

Security concerns served as the perfect cover for some extraordinary sharp practice. Before legitimate suppliers even knew the Kenya government was looking for a particular service, news would break that the deal had been signed. To experts in the field, the contract awarded might seem suspiciously expensive – sure sign of a 'commission' being paid – but since the government's contract specifications were not available, it was impossible to tell how much was being creamed off. 'In the absence of competitive bidding, it was not possible to ascertain how the contract sums were determined and accepted by Government as fair,' Kenya's auditor general, Evan Mwai, would note in a damning April 2006 report into Anglo Leasing.[27] One small example, however, gave a hint of the level of greed involved: Kenya was paying $US9 million each for MI 17 helicopters; a quick internet trawl revealed them to be simultaneously selling in Asia for just $3.9 million.

No due diligence had been carried out, the auditor general discovered, no implementation schedules agreed. In some cases, the companies concerned might actually have planned to supply the Kenyan government with promised equipment, albeit at ridiculously inflated prices, but once again, noted the auditor general, 'the non-availability in most cases of detailed contract specifications, invoices and delivery notes' made it difficult to verify what was delivered, and the failure to keep an up-to-date register of assets left their whereabouts unclear. In others, the Kenyan government clearly never stood to receive anything at all, given the nature of the companies it was doing business with. 'At least seven of the supplier/credit providers do not exist in the countries in which they are purportedly registered and may therefore not be bona fide registered business firms,' Mwai found. 'Additional firms among the list of the suppliers/credit providers may also prove to be non-existent.'

Confusion over the firms' identities was an important ingredient in the scam. As John probed the eighteen contracts the Kenyan media and public would come to refer to collectively as 'Anglo Leasing', he would be confronted by the surreal situation of a government which appeared to have no clear idea with whom it was signing its multi-million-dollar deals. Just as my office manager prevented me from identifying his fellow conspirators by inventing a shadowy door-to-door salesman, the anonymity of the contractors helped conceal both the mechanism and eventual beneficiaries of this laziest of scams. It was always obvious that the identities of the entrepreneurs involved in Anglo Leasing could not be the great mystery claimed by the ministers and civil servants involved. How could they *not* know their suppliers, given the supposed sensitivity of the projects? How could they later go on to cash cheques from these nameless individuals, refunding monies paid? The very questions were absurd, yet no one in power was in any rush to ink in a deliberately created void. No wonder newspaper cartoonists drew Anglo Leasing as a featureless ghost, a figure with the sizeable belly, business suit, bulging briefcase and bloated potato shape of one of Kenya's *wabenzi*, but without nose, eyes or mouth.

There was another fishy thing about Anglo Leasing: the payment methods. These were all 'turnkey' deals, in which suppliers offered not only equipment, but the funding arrangements a cash-pressed African government needed to pay for all this state-of-the-art hardware. The government signed a 'credit supplier contract', under which it was loaned the money, undertaking to repay the credit via irrevocable promissory notes.

In the Moi era, when the IMF and the World Bank grew wary of funding Kenya, such special credit arrangements had a certain logic. But by the first year of the NARC regime, Kenya was receiving over half a billion dollars in aid from its foreign partners, much of it at interest of less than 1 per cent. The interest charged on the eighteen Anglo Leasing contracts, Mwai would later establish, ranged between 4 and 6 per cent. Despite its ready access to cheap money, the government preferred to borrow at a hefty premium. As for the use of irrevocable promissory notes, the practice simply stank.[28] Backed by a legal opinion from attorney general Amos Wako, these notes were eternally binding. As good as cash, they could be bought and sold on international financial markets. If a Kenyan government subsequently tried disavowing them, it would do so at the price of its creditworthiness and its reputation.

On an infrastructural project of any size, cautious customers usually insist on paying in tranches, checking that targets have been met and a certain quantity of equipment delivered before handing over the next cheque. Not so with the Anglo Leasing deals, in which the government displayed a baffling eagerness to cough up, even initiating payment before the projects had actually begun. 'In some cases, the full repayments of the credits were accomplished before the projects were completed,' wrote Mwai, logging one case in which the loan was fully repaid five months ahead of schedule.

At every step, the government seemed to place suppliers' interests before its own. And all for what, exactly? In its talks with foreign donors, the government routinely put education, the fight against malaria, the digging of wells and the fight against AIDS at the top of its agenda. These were surely the correct priorities in a developing African country where nearly half the population lived below the poverty line. Yet now, it seemed, one of the world's poorest governments deemed digital communications and state-of-the-art surveillance equipment so vital it was willing to put future Kenyan generations in hock to secure them. As the British high commissioner Edward Clay put it: 'Why bid for all these sophisticated computerised programmes when your own official government strategy puts computerising far below things like anti-measles vaccinations?'

It didn't take a genius to work out what was really going on. The Anglo Leasing contracts were a crude device for extracting large wads of money from the Kenyan Treasury. Where the funds would eventually end up was anyone's guess, but it was safe to assume they would be split between those in government who authorised the deals and the entrepreneurs who provided the necessary camouflage by setting up a range of respectable-sounding shell companies and credit providers – 'looting pipes', John called them. The hope must have been that the confidentiality of these security contracts would muddy the waters long enough for the perpetrators to get away scot-free. Sometimes the suppliers at the end of the chain actually existed, although one had to wonder why any legitimate firm would agree to become embroiled in such intricate, shady deals. In other cases, the supplier was no more than a street address. Twelve of these contracts had actually been signed – if not activated – under KANU and six under NARC, a detail which highlighted one of the most intriguing aspects of the scam. This was an apolitical money-making venture. The shady players who had originally sold the idea to the Moi administration had not let the change in regime put them off their stride. Supremely flexible, these amoral, shape-shifting pragmatists had simply made their pitch to State House's new incumbents.

The fact that many of these middlemen were of Asian origin had its roots in the country's colonial history. The white administration brought indentured coolies from the Indian subcontinent to Kenya to build the railway. Most returned to the subcontinent, taking word of the job opportunities opening up in East Africa, and their places were taken by skilled clerks, craftsmen and traders from the Punjab, Gujarat and Goa. The colonial authorities did not want these new arrivals buying land – that was for whites – and once Kenyatta Africanised the civil service, jobs in the public sector vanished. The one sector in which Asians could flourish unhindered was trade. From modest *dukas* in one-road towns, vast business empires grew, and with them a reputation for deals clinched with a nod and a wink. The emerging African elite felt little affection for the *wahindi*, seen as tight-fisted, snooty and brazenly colour-conscious, but its members knew they needed its business nous. With extended families stretching across half the world, only the Asians had the international contacts and backup to help a minister wanting to stash illegally-acquired funds abroad. Only they understood what legal hoops had to be jumped through to manipulate the import licensing process, establish a fraudulent bank or set up a shell company in the Canary Islands. Moi's Kalenjin, in particular, relied on Asian wheeler dealers like Ketan Somaia and Kamlesh Pattni – real-life models for the oleaginous anti-hero immortalised in M.G. Vassanji's novels – to hatch the strategies that would allow them to make their margins and hide the proceeds.

The key question for John was: where did the president stand in all of this? John's best-connected informants were painting a picture of a vacillating head of state, being pulled in one direction by the Mount Kenya Mafia and in the other by his conscience. John clung to that scenario. His boss might be weak, he told himself, but he was fundamentally decent. And if a war of influence was taking place inside State House, then John, the persuasive charmer, was determined to win it. 'I was advised to move even faster, as the political sharks were rounding against me, and I had to get to the president before they did.'

The fuss John had whipped up over the first two Anglo Leasing contracts was producing results. On 11 June, finance minister David Mwiraria stepped into his office to announce that Anglo Leasing had returned all the money it had been paid on the forensic laboratories contract. He was joined by justice minister Kiraitu Murungi, and together the two men urged John to back off, insisting that this was what the president had requested and warning that if he continued pursuing the case, the government was likely to fall, so many of 'our people' were involved. Four days later, John learnt that Infotalent Ltd, a company which had won a police security contract, had returned 5.2 million euros. By the end of June, almost a billion shillings had been repaid by bogus companies on the eighteen-item list. Some frantic back-pedalling was taking place.

It did nothing to allay John's fears. The civil servants concerned claimed to have no idea who was returning all this money: that blank-faced ghost making his appearance once again. Yet, in a moment of indiscretion, finance minister Mwiraria let slip that he had arranged Anglo Leasing's refund by instructing a member of staff to call Asian businessman Deepak Kamani. If Mwiraria already knew Kamani was behind Anglo Leasing, it made a mockery of the investigations being conducted by the KACC and the auditor general. The same haziness hovered around the figure of Merlyn Kettering, an American consultant whose name kept surfacing in connection with the eighteen deals. John's informers told him Kettering attended high-level meetings in the office of the president at which sensitive military and communications projects were discussed. Yet Dave Mwangi, permanent secretary for internal security, denied Kettering's involvement when quizzed in front of the president.

As he stumbled on lie after lie, John continued briefing the president on what he was learning. On the morning of 18 June, noting that

Kibaki seemed in high spirits, John decided the time had come to make his pitch. Circumstantial evidence kept pointing to the same players, he told the president over breakfast. Given the shambolic nature of Kenya's judicial system, the matter could not safely be left to the law. A political gesture was necessary; heads must roll.

He had overreached himself. Looking across at the man he admired, John caught an expression he had never seen before: Kibaki seemed, well, *sheepish.* Like a boy caught with his hand in a biscuit tin. The president urged John to slow down. Above all, he was not to hand the files he was compiling on the roles played by the two permanent secretaries to the attorney general. 'In essence, I was stopped. I had been put on ice,' he recorded in his diary: 'The war against corruption is in State House, and I have lost the president's support. H.E. has let me down.' He had entered the room full of energy. He left it with his weightlifter's shoulders slouched, his morale in his shoes. It was time to go, he told himself.

He was beginning to feel unbearably dirtied. He had learnt so much about Anglo Leasing, who was behind it, the sums involved. He was the anti-corruption chief, appointed to protect the Kenyan people from predatory politicians, yet in his own eyes he was virtually sitting on his hands, smiling, chatting and cracking jokes with the looters in the interests of keeping the peace. 'I was complicit. There was no doubt, I was complicit.' For someone who had imbibed the Opus Dei lesson that work is a means to sanctification, professionalism a form of godliness, nothing could have been more abhorrent. Sometimes he hated himself.

In John's mind, Kibaki was no longer an inspiring abstraction. He had become a man whose personal qualities – or rather, failings – were of huge, immediate significance. But one could swiftly drive oneself crazy trying to work out who the president really was, or what he genuinely believed. Kibaki's survival, John increasingly realised, was based on his very amorphousness. 'Because he's so enigmatic, people see in him what they want to see. People will tell you, "He's incredibly wishy-washy," others say, "He's very indecisive," others say, "He's actually very cruel." He's everything to all men. I can't say I was immune.' Craving a reformer, he had persuaded himself Kibaki was that man.

Yet John stayed his hand. Concern for his family was one factor. His father had been almost financially ruined when he had fallen out of political favour once before, and John had no desire to force him to repeat the experience. The old man was not in the best of health, nor was John's mother: they had reached an age where they deserved some peace of mind. His brothers and sisters were all building their lives, starting young families, moving into new apartments. If he fell from grace, if the Githongo family name became politically toxic, how would it affect the people he loved?

In Kenya, as in most African nations, the moneyed, well-educated upper class forms a numerically tiny group. The political elite, business elite and social elite are one and the same thing. Rubbing up against one another at private schools, in clubs and at high society weddings, its members share an incestuous intimacy. 'In Kenya,' one young woman explained to me at a dinner party, 'it's not so much six degrees of separation, as one and a half.'

If a sociologist were to try to capture the various *milieux* featured in this book in the form of a Venn diagram, he would find the circles overlapping each other so heavily the categories became almost indistinguishable. One large circle, labelled 'St Mary's school', would include John himself, presidential aide Alfred Getonga, wheeler dealer Jimmy Wanjigi, anti-corruption campaigner Mwalimu Mati, opposition leader Uhuru Kenyatta and David Kibaki, the president's son. Another circle, intersecting the first in unexpected ways, would be labelled 'Received medical treatment from Dr Dan Gikonyo'; it would include John, his father Joe, Kibaki and Nobel Prize-winner Wangari Maathai. A circle branded 'Membership of Muthaiga, Karen, Limuru, Kiambu and Nyeri golf clubs' – Kibaki; TI board member and University of Nairobi chancellor Joe Wanjui; George Muhoho, head of the Kenya Airports Authority; Mateere Keriri, former State House comptroller; future defence and transport ministers Njenga Karume and John Michuki – would take a huge bite out of the circle marked 'Democratic Party founding members', and would virtually swallow up two others labelled 'TI-Kenya Board' and 'GEMA'. A circle labelled 'Practising Catholics' would overlap with all the smaller circles. All this before the circles for ethnicity were even drawn.

It was all very cosy. When things went well, of course, these networks were a great source of strength, a safety net stretched out in anticipation of life's shocks and reverses. But for anyone out of tune with the times, each link felt like one of the slender ropes the tiny citizens of Lilliput used to tether the giant Gulliver to the ground. As soon as John tried to lift an arm or raise a foot, he became aware of a delicate cobweb of expectations, obligations and duties tying him down.

As June ticked by, his relations with colleagues grew ever more strained.

Listening to the recordings John taped around this period, what's jarring is the laughter. Missing the undercurrents, unable to see the darting eyes and uneasy body language, an outsider would be forgiven for assuming these are old pals having a whale of a time. Underhand methods are explored, violent outcomes hinted at to a steady chorus of Santa-Claus-like 'ho, ho, ho's. Blackmail may be attempted, death threats pronounced, but anyone would think it was all some great, back-slapping joke, a delightful exercise in male bonding. A friend of John's would instantly have recognised these as the baritone chuckles John produced when he was nervous or mentally on the run, a world away from real humour. But it was laughter nonetheless.

'I would always respond by trying to make a joke of it,' acknowledges John. 'It was the only way. If you fell silent and the room went quiet, then . . .' He pauses.

'Then what?' I asked.

'Then you would have to deal with the uncomfortable realities as they had just been presented.' To name something is to allot it its rightful place in the universe, imbuing it with power. Not naming allows a measure of ambiguity – that necessary Anglo Leasing ingredient – to be retained. John came to label it 'The Culture of the Deadly Smile'. Al Capone and his lieutenants, Caligula and his aides, must have had just such strained exchanges as these, where the fake smiles constantly trembled on the verge of full-throated snarls.

Conciliatory by nature, John was never the type for the macho standoff. Yet he became embroiled in a one-hour argument with the head of the civil service, Francis Muthaura, that escalated into a shouting match. Muthaura accused John of leaking Anglo Leasing stories to the media, and insisted on putting out a reassuring press release that John regarded as so misleading it constituted lying to the Kenyan people. A call from Philip Murgor, director of public prosecutions, alerted him to the fact that Alfred Getonga was asking whether a batch of letters John had arranged to be sent to foreign banks, seeking to establish the identities of the shadowy figures wiring money back to Kenya, could be recalled. Far from persuading the president to sacrifice his tainted aides, John now heard rumours that he was the one about to be moved. On 29 June, without advertising the fact, he surrendered his diplomatic passport and took out a new one, allowing him to travel as an ordinary Kenyan citizen. He also had his frankest ever exchange with Kiraitu. 'He said it was now clear that Anglo Leasing was "us" – our people. He said no matter what, he did not have what it took to order or countenance the arrest of Chris Murungaru for corruption because they had too much history. He was blunt and emotional,' John wrote in his diary. 'You are conducting the fight against corruption like a person burning down a house to kill a rat,' Kiraitu admonished him. John shrugged. 'Killing rats is always a damaging business,' he replied. The justice minister's closing words were pointed. 'Tomorrow,' he warned, 'is the kind of day reshuffles happen.'

And so it proved. The atmosphere in State House the following day was charged. Senior officials came and went, slipping quietly into the president's office, exiting with hurried steps. Rumours buzzed from corridor to corridor, each contradicting the last. Rows of chairs were being lined up on the lawn in anticipation of a press conference, announced first for 13.00 hrs, then 14.00 hrs – but by nearly 16.00 hrs nothing had happened. John heard that both Murungaru and Kiraitu had been in the building, yet ominously neither had dropped by to see him. Finally, everyone scrambled: the television cameras were ready, journalists mustered, microphones switched on. The president mounted the podium and prepared to announce his new cabinet. In his office, John pulled up a seat alongside his staff and someone turned up the volume on the television. The names and titles rolled. No mention of John yet. On it went. And then, at the end, with the very last name, it came: John Githongo was being transferred to the ministry of justice. The axe had fallen. John had lost his precious access to the president, the favoured status which had made it possible to bypass ministers and gainsay permanent secretaries. He had lost Kibaki's ear, and would now answer to a minister who had made his position on Anglo Leasing abundantly clear. There was a shocked silence. John looked at Lisa Karanja.

'Well, that's it, then.'

'They have won,' said his secretary simply.

Immediately, John's mobile started buzzing with incoming text messages – some *faux*-sincere, some sardonic, others genuinely sympathetic – 'congratulating' him on his demotion. One call was from Charles Njonjo, former attorney general, once Moi's *éminence grise*, a man who knew what it was to fall from favour. His message was stiletto-sharp. 'You ARE going to resign, aren't you?' Another call was from John's brother Mugo. 'Have you been moved?' 'Yes.' 'Then you need to resign.'

There seemed no reason now to remain in State House, so the team decamped to John's place. That evening, old friends from his civil society days turned up to commiserate and the get-together turned into a spontaneous party. John had failed, and with the knowledge of his failure, he was surprised to discover, came not depression – not now, at least – but relief. The invisible arbiter to whom he mentally presented his conduct might conclude that he had made mistakes, but could never accuse him of bad faith. Anglo Leasing, the ugly and spectral guest at virtually every encounter of the last four months, was no longer his problem. It was true that in the middle of the laughter and music and chinking beer bottles, Stanley Murage, permanent secretary in charge of strategy at State House, rang to mumble something about John's relocation not being the 'real thing'. The idea was so preposterous John ignored it, and partied on. That night he slept through without waking, for the first time in months.

John spent the following day at the ministry of justice – the downtown skyscraper to which his emasculated operation was soon to be relocated – while his staff began packing boxes. It felt like an exercise in humiliation. The announcement of his demotion had triggered cries of alarm in the diplomatic community and civil society. Transparency International, his old employer, took out a full-page advertisement in the Kenyan press to protest, indicating that it would call off a planned international conference on corruption. But Kiraitu Murungi made no attempt to hide his glee as he inducted John into his new role. He was in full I-told-you-so mode. He had done all he could to save John, he claimed, but the combined forces of Murungaru, Getonga, Mwiraria and Muthaura, all pushing for John's ousting, had proved too much. Back at home that evening, John began planning his departure overseas with a friend. There seemed little point in staying on.

But the following morning he found Stanley Murage waiting for him in his State House office, insisting that John's demotion was not what it appeared. A member of the Mount Kenya Mafia, Murage claimed, had surreptitiously inserted John's name into the presidential speech. It was vital that he see Kibaki immediately. The tête-à-tête that followed was one of the most surreal of John's life.

When he told the president he had ordered his staff to start packing, Kibaki looked sincerely shocked.

Why he demanded, was John doing that? John replied that he had been transferred to the ministry of justice.

Who, Kibaki asked, had ordered that?

The ground felt as though it was shifting under John's feet. The order, he stammered, had come from His Excellency. He himself had watched Kibaki read the announcement out on live television.

Not possible, insisted the president. Looking genuinely upset, he summoned his justice minister to provide an explanation.

John stared at Kibaki, flabbergasted. He had thought the president virtually recovered from his stroke. Was it possible that the old man had been far more seriously affected than anyone had realised? He was not surprised that Kibaki did not write his own speeches. But was it really possible that he did not even take in the words that passed his lips? Or was this just a charade? Was General Coward playing his old game of trying to please everyone all of the time? Had the diplomatic and civic society reaction to John's removal been more negative than anticipated, prompting a rowback? Was a president immune to embarrassment actually using an infirmity to cover up a U-turn, preferring to be labelled senile than to seem soft on sleaze in foreign donors' eyes?

John would never know the answer. The entire baffling episode would, in retrospect, come to seem like a distillation of the nebulous, shape-shifting Kibaki presidency, where blame was passed from one player to another until the exhausted enquirer lost interest in the quest.

John could only wonder, and enjoy the spectacle of Kiraitu, so smug the day before, being humiliated in his turn as an irate Kibaki informed him there would be no change. The following day, the old routine nominally re-established itself, with John briefing the president on who he believed were the key players in Anglo Leasing, and Kibaki urging further investigations, on the naval frigate contract in particular. But the trust had gone. Quietly, John began putting feelers out to friends in the NGO and Western academic worlds, delicately exploring what avenues would be open to him should he jump ship. Many of his informants quietly went to ground following this episode, and some never resurfaced. For them, the post of anti-corruption czar had lost its aura of invincibility. They registered something John did not want to acknowledge. Looking back, he would come to date his abortive demotion, The Presidential Announcement That Never Was, as the moment any hope of a genuine crackdown on graft quietly died. His office remained in State

House, and the friendly chats with the *Mzee* continued, but something in their relationship had changed for ever.

John had fallen victim to one of the continent's oft-rehearsed myths. Reporting Africa, I've always been puzzled by the readiness otherwise intelligent diplomats, businessmen and technocrats show in embracing the 'Blame the Entourage' line of argument. 'The Old Man himself is OK,' runs this refrain, echoed at various times from Guinea to Ivory Coast, Zaire to Gabon, Tanzania to Zambia. 'Deeply principled, a devout Muslim/Protestant/Catholic, he observes, in his own life, a strict moral code. It's his aides/wife/sons who are the problem. They're like leeches. If only he'd realise what they are doing in his name and put a stop to it. But of course he adores them. It's his one weakness. Such a shame.' The argument has always struck me as a form of naïvety so extreme it verges on intellectual dishonesty. In countries where presidents have done their best to centralise power, altering constitutions, winning over the army and emasculating the judiciary, the notion that key decisions can be taken without their approval is laughable. If a leader is surrounded by shifty, money-grabbing aides and family members, it's because he likes it that way. These are the people he feels at ease with, whose working methods he respects. Far from being an aberration, the entourage is a faithful expression of the autocrat's own proclivities.

Despite his journalistic scepticism, his written musings on this very subject, John had swallowed the line. After seeing Kibaki in his hospital bed watching TV cartoons, he had convinced himself that if State House was running off the rails, the blame rested with a coterie which had taken advantage of an upstanding man whose wits were temporarily scrambled. The six months following Kibaki's hospitalisation, he initially told himself, had been a lacuna during which all sorts of underhand activity had flourished uncontrolled. 'I would think, "He's in it," but then, of course, he had been ill . . .' In fact, John was now forced to admit, the opposite was true. Dining regularly with the president, he could track his steadily improving mental health as accurately as anyone bar Kibaki's doctor. 'I was eating with him and I had a very good read on things. He was improving steadily, he could talk about something today and discuss it accurately tomorrow. I would sit back quietly at home, look at it rationally and say: "Actually, the fitter he gets, the worse this problem has become, the more confident the crooks are."'

11

Gorging Their Fill

'He's a menace to the donors, he's the taxpayers' despair,
For the Treasury is empty – but Macavity's not there.'

British high commissioner EDWARD CLAY's bastardised version of 'Macavity: The Mystery Cat'

John's struggles were not going unobserved outside the confines of State House. Watching from the sidelines were the media, civil society and Western donors. This last group had a particularly keen interest in NARC's performance on corruption. Having piled into the country with promises of aid when Moi quit the scene, they needed reassurance that it was not sliding back into the bad old ways.

Kenya is one of a raft of African nations locked in a symbiotic – perhaps 'mutually parasitic' is a more accurate term – relationship with the developed world and the lending institutions set up at the Bretton Woods conference in the wake of the Second World War to combat global poverty. Post-independence, its agricultural sector received British support, but it was with the oil shocks of the 1970s and the collapse of commodity prices that Kenya's economy really began to depend heavily on loans, grants and investment from an industrialised world ready, as the Cold War locked Africa in its icy grip, to provide 'no questions asked' funding to any government rebuffing the Soviet Union and the Communist bloc.

Seen as too important to be allowed to fail, Kenya became the first sub-Saharan country in the 1980s to receive structural adjustment funding from the IMF. Between 1970 and 2006, this modest African country received a total of $US17.26 billion from its foreign allies, roughly one and a quarter times what the Americans spent on the Marshall Plan, designed to rescue the whole of war-ravaged Europe. At its height in the early 1990s, aid from both the multinational lending institutions and donor nations which followed their lead accounted for 45 per cent of the Kenyan government budget. Under Kibaki, that shrank to less than 5 per cent, thanks to improved tax collection, but the total – $768 million in 2005 – remained hefty by any standards.

The relationship between giver and receiver has rarely been free from strain. In the early years, donors carefully steered clear of what were known as 'governance issues', a polite euphemism for 'graft'. Raising the issue in the prickly era when memories of colonial injustices were still sharp was deemed intolerable interference in a nation's sovereign affairs. The World Bank and the IMF's *raison d'être*, their members argued, was to fight poverty, not corruption – which was a political, not an economic issue. The cynical realities of the Cold War required loyalty payments, and if those had to be made to military dictators and autocrats with blood on their hands in the knowledge that little would reach the poor, so be it. But with the passage of time came the growing realisation that financial transparency, human rights and institutional checks and balances mattered more to the quest for prosperity than had previously been recognised. Africa's insatiable Big Men were in danger of killing their own economies, and as crisis bit, their pillaging began to hurt more than it once had.

Kenya perfectly illustrated that shift. Kenyatta had undoubtedly been both authoritarian and corrupt, but a healthier economy meant the looting drew less attention. Moi's far leaner economy struggled, in contrast, to sustain the impact of State House's system of authorised looting, which a minister later estimated to have cost the taxpayer a total of 635 billion Kenya shillings (roughly $US10 billion) in the space of twenty-four years.[29] Identical practices were viewed in the West with freshly critical eyes. The fall of the Berlin Wall meant support for disreputable African regimes could no longer be justified on traditional 'He may be a bastard, but he's *our* bastard' lines. Many of those bastards were seen to produce failed states, judged – with the menace of the Soviet empire gone – to threaten Western interests. The new head of the World Bank, the Australian James Wolfensohn, gave voice to the new approach when he declared corruption an 'intolerable cancer'. Donors began digging in their heels, insisting on elections and attaching ever more detailed conditions to their money. Wily presidents responded by playing games. That tension explained a pronounced stop-start pattern to Kenya's aid flows. Jerky as a car driven by a learner with no clutch control, relations between the two sides revved, stalled, and then sputtered back into life as trust came and went.

Like other African strongmen before him, Daniel arap Moi proved a master in running rings around the donors. Economists use the term 'reactance' to describe leaders' tendency to demonstrate their autonomy by doing precisely the opposite of what their Western supporters want. But Moi's contrariness was not simply a form of bloody-mindedness. Only a fool could have failed to spot that most latter-day World Bank and IMF reforms implied a radical trimming of emperor-like powers. At times in Kenya the donors appeared to carry the day, as happened in 1991 when, exasperated by Moi's refusal to allow multi-party politics, they suspended $350 million-worth of aid. Moi responded within weeks, amending the constitution to legalise opposition parties. At others, Moi emerged the victor by dint of sheer tenacity. He would fiercely resist a suggested change throughout months of negotiations, then appear to give way, only to implement it in a way that made a mockery of the entire exercise, effectively sending a two-finger salute to his foreign partners. Economist Paul Collier logged how the Kenyan government promised the same reform – of its maize marketing system – to the World Bank at least five times over a fifteen-year period in return for aid, only to reverse it on each occasion. 'The amazing thing is that the money kept coming. How did Kenyan government officials manage to keep straight, sincere faces as for the fifth time they made the same commitment? How did officials of the agency manage to delude themselves into thinking that adherence this time was likely?'[30]

It was a similar story with the establishment of the Kenya Anti-Corruption Authority (KACA). Aware that both Kenya's judiciary and police force were rotten to the core, Western donors seized on the idea of setting up a special anti-sleaze body, a Kenyan version of Chicago's Untouchables. Its director would enjoy security of tenure and its staff be appointed by an independent committee, shielding them from bribery and threats. Moi fought the proposal all the way, only caving in after the donors suspended $400 million in aid in 1997. He then appointed one of Kenya's most insalubrious businessmen, a former police officer with a terrifying reputation, as KACA's first director, compromising it from the outset. Six months later the director was suspended for incompetence, and in 2000 the high court declared KACA unconstitutional.

Western donors were not the only ones made to look gullible by the 'Professor of Politics'. In 1999, when the second major aid freeze was beginning to bite, Moi appointed Dr Richard Leakey, one of the few white Kenyans engaged in politics, to lead a 'Dream Team' of technocrats which would overhaul the country's nepotistic civil service. The president was desperate to get a new IMF lending programme and Leakey, no political naïf, became convinced he had finally accepted the need for change. 'When I went to see him, he told me to prosecute his own son, Philip, if it proved necessary. When you have that kind of assurance, you feel you can blast your way through.' The Dream Team worked wonders for a year. An impressed IMF, convinced Kenya was finally on the right track, approved a new agreement. 'A week after that, direct access to Moi was closed off,' remembers a rueful Leakey. With the removal of the president's blessing, the Dream Team hit the buffers. 'He'd got what he wanted. We had shot ourselves in the foot by securing that IMF deal far too quickly.' Leakey stepped down, aware he had been thoroughly outmanoeuvred. Moi might be led to water, but could only rarely be made to drink.

That legacy of presidential jousting and international distrust goes so deep that many of Kenya's donors – the EU is a notable exception – still steer clear of direct budgetary support, in which money is paid straight into government coffers. The equivalent of handing a beggar a fistful of cash and saying, 'Do with it as you see fit,' this is the most cost-effective way of funnelling aid, but it requires a reliable partner. The alternative, project support – the equivalent of telling the beggar, 'You can't be trusted not to waste your money on beer, so here's a voucher for a cup of coffee' – requires the involvement of NGOs and constant monitoring by donor officials.

Covered in detail in the Kenyan media, the fraught nature of Moi's dealings with the donors established a principle in the minds of ordinary Kenyans, understandably sceptical of their MPs' readiness to stand up for the ordinary citizen. If things got too bad, they told themselves, if the GSU was out in force and teargas billowing along Uhuru Avenue, the West would express its disapproval in terms a materialistic government understood: money. Hence the intense interest, bordering on obsession, in what foreign countries thought of the way Kenya was being run. As a journalist based in Nairobi I would often be taken aback by the way in which a routine story of mine, echoing what had already been stated *ad nauseam* in the local press, would be seized upon and reprinted for the benefit of the Kenyan domestic market. 'Britain's *Financial Times* Says Kenyan Economy on the Slide', the newspapers would trumpet, as though here was the final, ultimate proof that the government was doing a lousy job. A foreigner was saying it, so it *must* be true. Western donor governments, their media and their expatriates, had become the ultimate, trusted arbiters of Kenyan reality.

The truth was a lot murkier, the donors' stance rather less heroic than it at first appeared. The very factors that made them fret about governance in Kenya – the country's position as East Africa's dominant economy, its history as a capitalist ally, its military agreements – also encouraged a conviction that a certain amount of abuse was tolerable in the name of *realpolitik*. 'The British in particular have long held the quietly racist, patronising view that Kenyan affairs are being managed as well as anyone could expect, that the present government is the best we can hope for – in other words, that Africans simply don't have the intelligence or sophistication to manage very well,' John noted in *Executive* in February 1994. There was a tendency to favour democracy's concrete symbols – election day – over its substance, which required constituencies to be redrawn and the constitution changed if the opposition was to have a fighting chance. Rigged elections were hailed as 'steps in the right direction', a euphemism for 'good enough for Africa'. When, after the 1997 elections, the donors found that KANU victories in eight constituencies did not stand up to scrutiny, a finding which cancelled out the ruling party's parliamentary majority, they agreed not to mention this awkward fact in their final report on the polls. 'The donors' primary concern appeared to be the avoidance of any path that could lead to a breakdown of the political and economic order,' wrote academic Stephen Brown. 'Fearing instability, looking for quick results and avoiding more uncertain but farther-reaching reforms, donors actually forestalled more fundamental change.'[31]

Britain wanted to continue training its soldiers in Kenya. Washington, in the wake of the 1998 bombing of the US embassy in Nairobi and attacks on Israeli targets on the Mombasa coast, wanted a pro-Western administration in power in Kenya, a bulwark against Islamic extremism in the Horn of Africa. In a region which had seen both civil wars and genocides, there was no point being too purist. In their defence, foreign officials could point to a string of development studies showing that to be effective, aid had to be disbursed consistently, not turned on and off like a tap as punishment or reward. To make the most of Western generosity, African finance ministers needed to know how much they could expect in five or ten years' time. For the sake of the poorest of the poor, they argued, donors must lift their eyes from the petty abuses of the day and take the long view. Institutions and the moral imperatives they enshrined took decades to establish, checks and balances years to shore up, civil society a lifetime to build. To their credit, the donors contributed funds to all three processes, footing the bill for prosecutors' offices to be revamped, technocrats hired, media conferences staged and human rights groups established. The harvest would be reaped a long way down the line. Genuine concern required a curious form of brutality, in which the gaze was trained, Buddha-like, on the far horizon.

Less creditably, another innate human tendency was also at work. In every enterprise there is a powerful urge – particularly pronounced in those of a methodical, paper-pushing nature – to Keep the Show on the Road. Habit creates its own compulsion, supposedly temporary projects develop a momentum and logic of their own, and the larger the sums of money and the staff numbers involved, the harder it is to admit a thing has run its course. The inclination to Carry on Carrying On means donors shrink from walking away, no matter how disappointing the results of their intervention.

At an individual level, there were also solid professional reasons encouraging lenders not to rock the boat. In any development organisation, whether USAID or Oxfam, the World Bank or World Vision, career progress is measured in how much money an official succeeds in 'pushing out the door'. No one gets Brownie points back at head office for closing down a programme or putting a relationship with a client government on ice, even if this was, in fact, the most constructive course of action. Humanitarian organisations may talk about making themselves redundant, but their annual reports rarely boast about offices closed or staff laid off. Organisations' internalised incentives all work in the opposite direction, and the short stints on which careers are structured also make it all the easier to err on the side of indulgence, militating against the build-up of institutional memory. Fired with enthusiasm, an arriving director is baffled to discover how few projects his predecessor approved. The man must have been either lazy or lacking in vision, he assumes, and sets to work with a will, only to start encountering the same sharp practices that demotivated his colleague. At precisely the moment when he has reached a mature understanding of just how formidable the system he hopes to reform really is, the director is pulled out. As the East Africa correspondent for the *Financial Times* – itself a short-stint job

– even I stayed long enough to be able to accurately predict, purely on the basis of how far on in their postings they had advanced, how naïve or sceptical the donor representative I was due to interview was likely to be.

Finally, there were social and cultural factors working in favour of maintaining the status quo. World Bank directors are hardly average citizens. Often government ministers in their former lives, they belong to the international elite that automatically turns left on entering a plane, rarely does its own driving, and expects lodgings to come with staff quarters attached. The fact that those assigned to Africa usually themselves come from the developing world is of little relevance: these men – and they are almost always men – did not grow up in the local equivalent of Kibera. If they had, one suspects their view of the government of the day would be somewhat more jaundiced and confrontational. As it is, most regard the African politicians and central bank governors they meet in wood-panelled offices to discuss the fate of the nation as social equals. Their children attend the same schools, their wives patronise the same hairdressers. Mixing on a daily basis with the great and the not-so-good, it is easy for them to assume the best of intentions on the part of their hosts. The World Bank's move in the 1990s to delegate more power to country directors, meant to democratise a top-heavy organisation, exacerbated a latent danger. Staffers showed an alarming tendency to lose their critical distance, over-identify with their clients and start regarding the interests of the local government and the World Bank as one and the same. 'To those with fragile egos, being seen as someone who has an "in" with the government matters a great deal,' says one World Bank veteran. Back at headquarters, they dubbed the syndrome 'governmentitis', or 'being captured' by a host state.

In Kenya, the full, unhealthy intimacy of that relationship was made apparent thanks to Lucy Kibaki, first lady and ageing drama queen, a woman widely feared for her hair-trigger temper and erratic outbursts. Late in the evening of 29 April 2005, a fracas broke out at the Muthaiga residence of Makhtar Diop, World Bank representative of the day. Diop, a former Senegalese finance minister, was throwing a leaving party to celebrate the end of his posting. Kenya's most popular musicians had been hired for the night, waiters were serving drinks, and the cream of Nairobi's political and diplomatic set was milling on the lawn when things began to go awry. Diop had reckoned without his next-door neighbour. Never one to wear her privileges lightly, Lucy Kibaki interpreted the loud music as a personal affront. Storming into the garden, she screamed at Diop's guests – who included two of her own children – that this was Muthaiga, not Korogocho slum, and tried to unplug the music system. She made three such commando raids before being barred by Diop's guards.

The episode stretched over several days, to the gathering glee of the Kenyan media. When journalists reported that Lucy had driven to her local police station to demand Diop's arrest, her fury found a new target. During a five-hour overnight scene at the Nation Media Group, staged in front of her mortified bodyguards, she harangued journalists for their disrespectful coverage of the Kibaki family, confiscating their mobile phones and slapping a cameraman in the face.

The footage of Lucy lunging at reporters played repeatedly on the television the following day, to shocked gasps and mocking smiles. But Kenyans so relished the sight of a first lady making a spectacle of herself, they virtually missed a fascinating detail exposed in the course of the spat, more worrying than all the diva dramatics. As it happened, Lucy had some justification for feeling events in Diop's garden directly concerned her. She was more than his next-door neighbour, she was his landlady. The fact that the World Bank representative was renting his house from the presidential couple – living in the very mansion the Kibakis had occupied before moving into another on the same plot – went some way towards explaining her behaviour. Why, the two properties were even linked by a little footpath. The official responsible for telling World Bank headquarters whether Kenya deserved millions of dollars in aid, who influenced donor governments into granting or withholding further millions in bilateral assistance, saw nothing amiss in paying rent to the president of a client nation. That moral myopia was not Diop's alone, for a team of World Bank auditors who visited Kenya three years before the party had found nothing inappropriate in the chumminess of his lodging arrangements. The failure to detect so much as a whiff of a conflict of interest spoke volumes about World Bank attitudes to those in power.

The entire episode highlighted the dangerous cosiness that can develop between two entities which should, ideally, be in a state of creative tension. In the mid-1990s, following the toppling of Zairean dictator Mobutu Sese Seko by a coalition of neighbouring African states, the donors seized upon the notion that a group of 'Renaissance leaders' – former rebel commanders turned progressive leaders – had emerged who would apply 'African solutions to African problems'. In the ensuing years, each of those leaders' credentials would become tarnished, but those who had engaged with the Renaissance leaders were loath to admit a mistake. It's hard to believe that influential American economist Jeffrey Sachs isn't mortified by his citing, in the introduction to his best-selling *The End of Poverty*, of Meles Zenawi of Ethiopia, Olusegun Obasanjo of Nigeria and Mwai Kibaki of Kenya, three countries whose elections saw major bloodshed and systematic vote-rigging, as members of a supposed 'new generation of democratic leaders' pointing the way ahead with their 'powerful and visionary leadership'. The wishful thinking is not exclusive to Sachs. Reading the websites and reports of organisations which lend to Africa, it's easy to log a series of tactful omissions, from the glossing over of security-force crackdowns to the editing out of eruptions of top-level 'misgovernance'. The small acts of self-censorship amount to a sympathetic retouching of a battered portrait.

A few years ago, a Kenyan beer company known for the patriotic slickness of its marketing campaigns came up with its most impressive offering yet. Set to orgasmic music, the TV advertisement showed a victory parade passing along central Nairobi's Kenyatta Avenue in a swirl of confetti. But this was not a Kenyatta Avenue any commuter would recognise. The *real* Kenyatta Avenue is a decrepit thoroughfare where glue-sniffing street boys and conmen mingle with harried office workers and Somali elders with hennaed beards. A stump-limbed amputee begs in the shade of a bedraggled hibiscus, a blind man holds out a cap, eyeballs rolling. The street's traffic islands have disintegrated into rubble, and when it rains pedestrians hop like grasshoppers to avoid being splashed by puddles of foul brown water. The advertisement showed the same urban skyline, but every face in the crowd was young and beautiful, every building immaculate, even the sky looked extra blue. How had they done it? I'd wonder every time I saw it. Had they gone out to film at 5 o'clock on a Sunday morning? Had the previous fortnight been spent painting the buildings? Or was this a tribute to a state-of-the-art editing suite? Whatever the answer, the advertisers had managed to conjure up a virtual reality Kenyatta Avenue, a thoroughfare which should exist, but sadly does not.

The donors' attitude to Kenya resembles that beer advertisement. There is the real Kenya, a poor land of glaring inequalities, and then there is the Platonic Ideal of Kenya, the if-only African state enjoying, in the wake of the 2002 elections, remarkably high growth rates and ever-improving school attendance, and whose 'progressive' leadership may occasionally fall prey to temptation but generally tries to do its best. That gleaming, shiny Kenya is the one the donors see in their discussions with government. As the beer ad illustrated, it is not necessary to tell an outright lie to convey an untruth, all it takes is a series of quiet omissions and small exaggerations.

By April 2004 the donors were seriously concerned. They had acted in concert over the introduction of multi-party politics, to startling effect, and they knew it was possible to do so again. The envoys of the United Kingdom, Germany, Japan, Canada and Scandinavian countries started meeting regularly at the British and US ambassadors' residences to air worries and share information. Everyone was free to throw in their tuppence ha'penny's worth. 'There were a number of streams coming in,' remembers US ambassador William Bellamy. 'It was a fertile exchange, everyone had something to put on the table.' John was invited to these encounters, but Bellamy remembers that when he did turn up he was 'not chatty, very circumspect', and always showed 'unwavering faith' in Kibaki. The diplomats joked about forming a 'Gang of Six' or 'Gang of Eight'. In fact, it was more a case of 'the two musketeers', for there were only two men with the staff numbers, in-house expertise and infrastructural backup to do a serious job of tracking unfolding events. The first was Bellamy: clipped, incisive, mustachioed. The second was Edward Clay, the British high commissioner.

Neither man was shy of airing his trenchant views, but there was a certain irony to the fact that Clay ended up with the far higher profile. Kenya's former colonial master is not the country's biggest donor these days: the United States overtook it long ago. What's more, British ambassadors are not the powerful figures of yesteryear. Since prime minister Tony Blair decided to separate development from foreign policy in 1997, the Foreign and Commonwealth Office has shrunk to a ghost of its former grandiose self, overshadowed by the ever-more-powerful Department for International Development (DfID), a thrusting giant with three times the budget. In many African capitals the head of the DfID office has more staff, a bigger fleet of gleaming white SUVs and larger premises than the ambassador, reduced to near figurehead status. But Clay's story would illustrate that tradition and myth – the myth that what a British high commissioner thinks of a former colony matters intensely – count more than the realities of economic and political clout.

When Clay first dared criticise the Kenyan powers that be, the irate foreign minister claimed such sentiments were only to be expected from the son of a colonial-era district officer. He was getting his Clays confused. Edward Clay is in fact the son of a former editor of the *Yorkshire Evening Post*, and very nearly became a journalist himself before veering off into the foreign service. The first time he set foot in Kenya was in 1970, as a twenty-five-year-old diplomat on his first overseas assignment. A lowly third secretary in Chancery, his job was to monitor Kenya's parliament and political scene.

First postings always leave a particularly strong impression. When Edward and Anne Clay returned to Nairobi three decades later, this time to occupy the high commissioner's residence in Muthaiga,

Kenya felt very much like home, despite the tripling in population that had taken place in the intervening years. In Kenya, British high commissioners have traditionally tended to be cut from emollient cloth. But there were several factors working to ensure that this particular incumbent did not fade quietly into the background. Clay was fifty-six, with retirement just four years off, and this was going to be his last posting, so in career terms he had nothing to lose. His youthful stint in Nairobi meant he barely needed a briefing to get up to speed. A useful network of friends and contacts – many of whom had now become influential politicians and businessmen – was just waiting to be dusted off. Then there was the character of the man himself. A boyish figure with a flop of grey-blond hair and a playful sense of humour, Clay made for the least stuffy and most approachable of ambassadors.

There is a type of resolutely silly humour at which clever English people of a certain class and education excel. Based on a keen appreciation of the absurd, it is associated with achingly bad puns, dirty limericks, quotations from the poems of Hilaire Belloc, vainglorious Latin mottoes, and a propensity for reciting long sections from *Alice in Wonderland*. Its whimsicality should never be mistaken for childishness, for it can mask ferociously serious intent, and the self-deprecation that often goes with it – another very English characteristic – is in fact a form of intellectual arrogance: so certain is the holder of his worth, he sees no need to force that understanding on others. Clay has that fey quality, and the misleading modesty that goes with it. Such individuals are hard to keep in line, for what they value – and mind losing – does not fit the conventional pattern. And Clay had already demonstrated his willingness to stick his head above the parapet before his arrival in Kenya, along with a knack for the florid turn of phrase.

An email received from a human rights campaigner Clay befriended during his second African posting, this time to Uganda, would come to seem ironic in the extreme. 'You did tell me and I still remember,' wrote the Ugandan, 'that I should avoid being confrontational because nobody would want to listen to what I have to say.'

Clay himself had been nothing like as discreet at this stage of life as the exchange suggested. As high commissioner in Kampala in the mid-1990s he had felt increasingly dismayed at the uncritical support London gave president Yoweri Museveni – one of the most high-profile and indulged of 'Renaissance' leaders – as he allowed a vicious guerrilla movement to devastate the northern third of his country, sending Ugandan troops instead to plunder neighbouring Congo. Such experiences suggested to Clay that donors did ordinary Africans few favours by constantly shifting the goalposts on what they considered 'acceptable' behaviour from aid-hungry leaders with no real intention of changing their ways. Frustrated by what he saw in Uganda, he spoke out on constitutional issues, helped shame a minister caught rigging an election into conceding victory to her rival, and repeatedly challenged Museveni over inflated defence spending. After he moved to Cyprus, his feistiness continued undiminished. When a Cypriot MP – a former pathologist – staged a seven-hour vigil on top of a British radio mast to protest against the presence of British bases on the island, the resident high commissioner dismissed him as a 'medical monkey up a stick', an expression that won almost as much attention as the protest itself.

Finally, like many of those who worked in the Great Lakes region in the 1990s, a horror story tugged at Clay's conscience, preventing him from relaxing into his last assignment. In Uganda he had also been responsible for Rwanda and Burundi, two small countries, he'd been confidently told by his Foreign Office bosses, he could expect to occupy just 1 per cent of his time. The Rwandan genocide, which began within weeks of Clay presenting his credentials in Kigali, made a mockery of their words. He was haunted by a feeling of undefined inadequacy in the face of the killings. 'If I'd been five years younger,' he said of his posting to Kenya, 'I wouldn't have had the confidence. I'd have felt constrained by lack of experience. But one thing I'd absorbed in my previous African posting was the costs of not speaking up.' Those in the Kenyan state apparatus who would later view Clay as the sinister hand of a Kikuyu-hostile British government missed the point just as thoroughly as they misunderstood John

Githongo's *raison d'être*: Edward Clay, by this stage of his life, had all that it took to become a rogue ambassador.

Arriving in Nairobi in December 2001, when John Githongo was TI's energetic young director, Clay noted with puzzlement the roller-coastering attitudes of diplomats posted to Kenya. 'In the first year, there was great enthusiasm: "We must increase aid." In the second year, revision set in. In the third year they all seemed to go bonkers, so disillusioned they couldn't speak or think rationally. I thought they'd all gone mad.' It never occurred to this diplomat, who would go down in history as one of the Kenyan government's most flamboyant critics, that his own attitude might trace a similar arc. Moi was on his way out after twenty-four years in which, whatever his failings, he had managed to keep the nation intact. So much seemed possible. This was a time to engage, not carp. 'There was a great feeling of a new departure, which is why I'd wanted the job.' And for quite a while Clay was happy to talk up a country he loved. When the US State Department issued a terrorism alert in December 2003 which threatened to devastate Kenya's tourism industry, already battered by a series of travel warnings, the local media learnt that Clay would be breakfasting at two of the hotels identified as possible terrorist targets, and would be available for photographs. It was a gesture of support appreciated by the Kenyan media, and he was briefly dubbed 'the two-breakfast diplomat' by the BBC.

But worries about terrorism, development and constitutional reform were soon subsumed by an overarching concern: Kibaki's loss of control. The comparison with his predecessor, a fitter man despite his extra seven years, was telling. 'Moi always had a grip on what was going on in the country and in his office. After Kibaki's accident, his authority fell into the hands of those around him. The sense of disarray was palpable. The donors were lining up, ready to engage, and they weren't getting any guidance.' As the vacuum persisted, stories of shady goings-on began surfacing. At first, nothing seemed connected. But as time went by, and connections between the various stories became apparent, an ominous pattern seemed to emerge.

Clay assigned a team at the High Commission the task of establishing what was going on. The Kenyan media, he felt, made the mistake of tackling each episode in isolation. The larger embassies could play their part, he believed, by putting together the pieces of the jigsaw to form a coherent picture. 'Those of us who had an institutional memory could stitch the various episodes together. We used to get all sorts of bits and pieces. There were three or four of us at the embassy working on it, and we did so continuously.' The team was small, effective and of like mind. 'We were all in tune with one another, and we all knew that if London didn't like what we were doing, Edward would take the rap,' recalls a former member.

The donor community was part-funding John Githongo's office, and also providing technical advice on constitutional affairs, so liaising with diplomats was part of John's official duties, something Kibaki had expressly asked him to do. But Edward Clay denies the oft-repeated claim that John spied for him. When the two met, they always chose highly visible spots, such as the Norfolk Hotel's exposed Delamere Terrace, to signal that there was nothing secret about their meetings. 'There were things we could advance to one another,' says Clay. 'But the idea that we exchanged files – that's a fantasy. John never betrayed anything he ought not to have done, and I don't think I did either. We knew our limits. It wasn't some kind of joint strategy. None of the diplomats were suborning civil servants.'

In fact, the diplomatic community didn't need to be leaked to. If Western surveillance in much of Africa can be a pretty sketchy affair, this is not the case in Kenya, regarded in London and Washington as too strategically important not to be closely monitored. Mobile phone conversations are worryingly easy to listen in on, and on a continent where the landlines have virtually collapsed, VIPs depend on their mobiles. The contents of those intercepts were fed to the key embassies, offering a fascinating insight into what Kenya's rich and powerful were up to. No wonder that when John met up with key diplomats, he was sometimes startled by the extent of their grasp of his patch. On one occasion during a meeting at the American embassy, a member of staff leaned over and said: 'John, we probably know more about this than you.'

Even without the intelligence, it was amazing what a systematic approach could yield – whether that meant checking company websites boasting about contracts with the Kenyan government, sifting through the rarely examined detail of the government's yearly budget or trawling old press cuttings to establish the track records of local entrepreneurs. That information could then be cross-referenced with snippets gathered via that most traditional of diplomatic techniques: old-fashioned schmoozing. Clay's resurrected network stood him in good stead. 'The classic diplomatic skills of getting out and talking to people really count for a lot,' he says. 'This is a very oral and anecdotal society. People will talk to you in a way they would never write to you. You could go around the ministries and report back on the mood music. There was a susurration of talk of corruption going on at many levels by people who were concerned.'

He reported what he was finding back to the other ambassadors. He also kept the World Bank's country director abreast, but the Kibakis' tenant, after an initial show of interest, manifested little appetite for what he was hearing. 'I let Makhtar Diop into quite a lot of it, who didn't believe me,' remembers Clay. 'And when he was forced to believe it, he declined to do anything about it.'

Clay felt supported in his digging by a new, more muscular line on corruption being touted in Whitehall. A growing recognition in the West that it took two to tango – for every minister trousering a bribe, there had to be a Western company ready to pay it – had culminated in the Organisation for Economic Cooperation and Development (OECD)'s Convention on Combating Bribery, signed by thirty-six member states. The convention became British law in 2002, for the first time giving British courts jurisdiction over crimes committed abroad by domestic companies. In America, bribing officials working for foreign governments is illegal, but that had not traditionally been the case in many European countries, including Britain. Now, thanks to the work of organisations like TI, a legally ambiguous era when foreign contracts were routinely clinched with massive 'sweeteners' paid to African *wabenzi* was supposedly coming to an end. Blair's government also gave its backing to the Extractive Industry Transparency Initiative, a scheme under which firms in the petroleum and mineral sectors agreed to open their arrangements with client governments to public scrutiny. So seriously did Whitehall take the new legislation, it sent a team to Kenya in 2004 to spell out its implications for the expatriate business community, and Clay and his colleagues gathered British and Kenyan businessmen together to warn them that corruption abroad could now lead to prosecution in Britain.

As the weeks passed and an increasingly unsavoury picture of bloated procurement emerged, the two leading ambassadors started taking their concerns to Kenyan ministers. Both of them assumed this could not represent government policy. For Clay, a meeting with the head of the civil service, Francis Muthaura, marked a personal turning point. 'I asked him what was going on. He made some Panglossian statement, assuring me that various contracts had been stopped, this was a dead horse we were flogging, and in any case it was a Moi-era horse. We had a real set-to. The colleague who was with me said he had never seen me so angry. By that time we had a pretty good idea there was a conspiracy abroad and that the various bits of malfeasance were linked. Some was old corruption being revived, some was new, but the constant theme was the networks were being recultivated and reactivated.'

Getting nowhere with the Kenyan government, Clay decided to go public, in a July 2004 speech to the British Business Association of Kenya. Circulated ahead of time to ensure no media outlet missed it, it was a typically whimsical Clay offering. Headlined 'Some Bread and Butter Questions', it ended with a pastiche of an old-fashioned children's rhyme. It was characteristic of the high commissioner that having got the serious stuff about corruption off his chest in the body of the speech, he should deliver a rambling spoof of 'The King's Breakfast', by A.A. Milne, inventor of Pooh Bear, fine-tuned during several enjoyable hours with his staff.

The King asked
The Queen and
The Queen asked
The Dairymaid:
'Could we have some butter for
The Royal slice of bread?'
The Dairymaid
She curtsied,
Thinking that this might be
A lovely little earner
If generously spread.'

The Dairymaid
Swiftly
Went and said to
The Alderney:
'About the butter contract
For the Royal slice of bread –
Are you interested
In long-term
Exclusive tendering?
For I heard that butter prices
Were about to leap ahead.'

The eleven subsequent verses sailed over most Kenyans' heads. But they had no problem grasping a few choice phrases carefully chosen by this journalist *manqué* to reverberate in a culture where power is so often expressed in terms of 'eating'. They were all the more shocking for being delivered in the clipped accent that has always allowed the Foreign Office's denizens to deliver the rudest of messages while remaining, in format at least, exquisitely polite.

'We never expected corruption to be vanquished overnight,' Clay told his audience. 'We all implicitly recognised that some would be carried over to the new era. We hoped it would not be rammed in our faces. But it has.' Those in government were now eating 'like gluttons' out of a combination of arrogance, greed and panic, he said. 'They may expect we shall not see, or notice, or will forgive them a bit of gluttony, but they can hardly expect us not to care when their gluttony causes them to vomit all over our shoes.'

'Vomit One', as Clay now refers to the speech, had an even greater impact than he had anticipated. He was summoned for a dressing-down by foreign minister Chirau Ali Mwakwere, and told to support his allegations with names, facts and figures. MPs accused Clay – embittered neo-colonialist that he clearly was – of insulting the Kenyan people. The newspaper cartoonist Gado brilliantly lampooned the government's hypocritical relationship with its foreign critics in the *Nation*, showing a drunken minister, vodka bottle in hand, vomiting copiously on Clay's feet. When Clay objects, the minister unleashes a torrent of belligerent abuse – 'Nobody tells me what to do in my own country, you hear me, so why don't you #*@* off?' – before suddenly holding out his hand. 'Er . . . Can you spare me a quid, mate?'

While recognising that Clay had expressed an ugly truth, the mainstream press was shocked by his indelicate language, riled that a diplomat representing Her Majesty's Government, with its brutal colonial history, should dare tick off a Kenyan administration in such terms. Even members of Clay's staff shared that view. 'With the benefit of hindsight, he went too far,' one told me. 'He was right, but in Africa, it's simply not acceptable to disrespect the Big Man, and that's what Clay was seen as doing.' In the gutter press, the smear machine began churning out its rocambolesque accusations. Clay was the president of the Royal Gay Society (leading member: one John Githongo) which staged weekend orgies in Lake Naivasha, scene of much aristocratic bed-hopping during the Happy Valley years. Clay's wife was in the pay of a foreign state, the planted stories claimed, and the high commissioner was so hated that the outraged Chris Murungaru, minister for national security, had tried to strangle him at a State House function.

But the reaction on the street was different. The British high commissioner's residence in Muthaiga lies at the end of an almost permanently traffic-clogged thoroughfare linking Nairobi's slums with its centre, so Clay had plenty of time on his way to work each morning to savour the public mood. Spotting his face – now familiar to every Kenyan – the touts selling newspapers at the traffic lights clustered round his car to urge him to greater efforts. 'You're right!', 'It was time!' *Matatu* drivers leant from their windows to cheekily offer lifts to Kamiti, Kenya's maximum security prison. Spotting the high commissioner's diplomatic plates, a policeman stepped forward to wave him through on a busy roundabout, a huge smile on his face. A former permanent secretary stopped him in the street, a twinkle in his eye, to remark that Clay's shoes looked remarkably clean. 'Thank you, you have done this country a singular service,' he added. When Clay visited his bank, the shoeshine men outside joined in the fun. 'Five shillings for shoeshine!' they yelled. 'Ten for vomit!' A beggar boy was spotted in the city centre with a sign reading, in Kiswahili: 'A penny please but don't puke on my foot'. At Nairobi's *nyama choma* joints, where roast meat was served with helpings of *ugali* maize meal, diners clasped bulging stomachs and joked, 'I'm so full, I could vomit on your shoes.' Ordinary Kenyans, it turned out, were rather less prone than the educated elite to post-colonial prickliness, bearing the brunt as they did of their government's predatory tactics.

Overwhelmingly positive, the public reaction came as a massive relief to Clay. In theory, his speech had been vetted by the Foreign Office, but staff in Whitehall, not expecting their envoy to Kenya to present a problem, probably skimmed it so quickly they missed the juicy bits. Clay knew he had caught his colleagues in London on the hop. 'It was clear to me that if it had gone wrong, I'd have been dropped like a stone. It was a bit like walking the plank. Instead, the British government saw what we saw, that there wasn't great public support for the official Kenyan line.' He had, in fact, become a hero to many Kenyans, in whose eyes he was playing the rambunctious role they expected of foreign donors. But Clay was about to run up against a far more formidable opponent than the Mount Kenya Mafia: his own government.

In public, British ministers supported their outspoken high commissioner. But as he set about drafting further speeches on government sleaze, Clay found he was fighting over words and phrases with Whitehall, fully alert now to their firebrand representative's capacity to stir things up. He was grateful for the lawyers' steers – no one wants to be sued for libel – but this went further. His colleagues in King Charles Street were essentially questioning the wisdom of speaking out at all. 'It was extremely tedious. I didn't expect London to try to teleguide what I was trying to do and say in my own way. I remember taking a call on my mobile from somebody in London trying to get me to pull one of my speeches as I was actually walking into the venue to deliver it. I said, "It's too late. And why?" He said, "We don't have time to go through all this," meaning: "We don't have time at the Foreign Office to have this internal debate." So I said: "Then leave it to me." By the time I gave Vomit Two, I'd fought over every word and sentence.' Clearing the speech took two months.

These conversations were with staff at the Foreign Office, but behind them Clay sensed a brooding presence. 'Someone was whispering in their ears.' That someone, he was certain, was the Department for International Development.

12

A Form of Mourning

'It isn't facing danger that cuts you up inside. It's the waiting, the not knowing what's coming.'

ELIOT NESS, *The Untouchables*

By the turn of the century, Western policy in the developing world was increasingly being set not in ministerial offices but by the NGOs – organisations like Oxfam, Save the Children, Christian Aid. The Make Poverty History campaign, pushing for the cancellation of Africa's foreign debt and dramatic increases in Western aid levels, was gathering momentum. Jeffrey Sachs, the brilliant American economist who campaigned in favour of a massive hike in funding, appeared to have won the emotional, if not the intellectual, argument. Other analysts might shake their heads at Sachs's simplistic formula for the continent's recovery, but he had successfully wooed pop-star campaigners like Bono and Sir Bob Geldof, and their ability to mobilise a younger generation bored by traditional politics awed Western governments. Whether on the right or left, political parties realised that promising to 'save' Africa was a potential vote-winner in the eyes of an idealistic coming generation. No wonder members of the African elite, aware of these pressures, sometimes sounded unappetisingly smug when contemplating tortured Western attitudes to the continent. As one Kenyan newspaper editor told me: 'What we Africans have realised is that your leaders need to lend to us more than we need to be lent to.'

An early convert to the cause of vastly boosted aid was Tony Blair, who denounced the state of the continent as 'a scar on the conscience of the nation' and called for a Marshall Plan for Africa. He pledged to raise British aid levels to the 0.7 per cent of GDP demanded by the NGOs and agreed to the creation of DfID, a ministry whose development agenda, separated from the Foreign Office, would no longer be tainted by national self-interest, so the theory went. The American-led 2003 invasion of Iraq, bitterly contested by the British public and deeply unpopular with the left wing of Blair's Labour Party, further boosted his interest in the continent. Africa was one of the few remaining areas where a compromised prime minister with a liking for the international stage could still show moral leadership.

In May 2004, just as the Anglo Leasing scandal was breaking, Blair launched the Africa Commission, whose recommendations – writing off debt, tripling aid by 2010 and improving trade terms – paved the way for the 2005 G8 meeting at Gleneagles in Scotland. The Gleneagles summit, Downing Street trumpeted, would be the climax of a Year of Africa, marking a turning point in the continent's relationship with the West. Playing to the industrialised world's guilt complex, the Make Poverty History campaign, Africa Commission and Gleneagles summit all shared one characteristic: the emphasis was on Western, rather than African, action. Top-down, statist, these initiatives were all about donor obligations, pledges and behaviour. What they definitely *weren't* about – despite token references to the importance of 'good governance' and a supposed pact between North and South – was highlighting the shortcomings of African governments set to benefit from future Western largesse.

DfID, the only British ministry with explicit instructions from the Treasury to boost rather than slash spending, swiftly won respect in other Western capitals for its focus, energy and principled insistence on untied aid. In development, at least, a fading imperial power could still claim to 'punch above its weight' on the global scene. At home, DfID's altruism, combined with its readiness to challenge decisions taken by fustier, more parochial departments, meant it was seen as a sexy place to work: 85 per cent of those applying to the Civil Service put it top of their list of preferences. Yet DfID's 'spend, spend, spend' philosophy was beset with difficulty.

Critics of international aid, like the American economist William Easterly, point out that one of the defining characteristics of the industry is its inherent unaccountability. An institution setting itself a narrow goal can statistically assess whether or not its efforts are having an impact. The more it takes on, the harder it is to separate out the various strands and quantify success and failure. 'Some of its goals are so huge as to be meaningless,' wrote *The Times* on the occasion of DfID's tenth anniversary. 'As well as saying that "our overall aim is to get rid of world poverty," it wants to scrap the EU's Common Agricultural Policy, complete the Doha Round of world trade talks and combat climate change.'[32] 'If you are responsible for everything, you are responsible for nothing,' comments Easterly. DfID gauges improvement by a country's progress in meeting the UN's eight millennium development goals, but DfID is not the only organisation disbursing aid, and recipient governments, after all, also play some role in determining their national course. This leaves the amount of money which is disbursed as the only solid yardstick of progress, hardly a situation likely to encourage discrimination amongst the officials who are responsible for approving projects.

DfID's first boss, the principled and forceful Clare Short, ruled that the department's efforts and funds should be focused on the 'poorest of the poor', countries where annual per capita income was $875 or below. This only seems right and proper. But nations emerging from civil war, authoritarian rule and military dictatorship are by definition those where governance is poorest, corruption rife and aid most likely to be diverted. 'Experience proves it's possible to usefully target aid, even when working in very corrupt environments,' DfID officials will tell you, insisting that projects are so carefully monitored in the field as to be effectively 'ring-fenced' from surrounding sleaze. Not possible, argues Daniel Kaufmann, former Director of Global Programmes at the World Bank Institute, who has spent decades quantifying graft's impact on development. 'The idea that donors can immunise their projects in a corrupt country is absurd, it's not what the evidence shows. When there is no integrity on the part of the leadership, no systematic approach to governance, civil liberties, rule of law, donor aid is simply wasted.'

The World Bank's own research bears out his scepticism, not just generally, but in the specific Kenyan context. A strictly confidential review by the bank's internal anti-corruption unit into four Kenya projects approved between 2000 and 2005, worth $375 million, found three suffered from 'serious irregularities'. It cited almost every imaginable stratagem for ripping off an externally-funded aid project, from the bribing of public officials to abuse of office, inflated expenses, fraudulent claims, conflict of interest, the concerted rigging of bids, failure to carry out allotted tasks, and blatant nepotism by MPs. Two of the projects were AIDS-related, and the report's compilers highlighted one of the most obscenely ironic consequences of the abuse: because grant money went to bribe officials rather than being spent on orphans' school fees, many children dropped out of education and resorted to prostitution. A project intended to reduce HIV infection helped, instead, to spread the virus.[33]

DfID, then, might be trying to do the impossible, seeking reliable partners just where they are least likely to be found. Even if these fundamental questions are put to one side and the correctness of DfID's philosophy accepted, problems of how to put it into practice remain.

One challenge is how to disburse the increasingly generous sums channelled DfID's way. In Africa, a shortage of qualified bureaucrats, lack of institutional experience and the absence of many of the legal checks and balances routinely required by Western partners mean poor countries struggle to access the money donors want them to have. Because of this 'low absorptive capacity', many African states regularly fail to claim the full amounts allocated them by Western governments each year.

Add to that the tarnished records of the Renaissance leaders the West once regarded as safe bets and it is clear DfID might struggle to spend its rising yearly budget. Having variously rigged elections, altered constitutions in order to hold office indefinitely, jailed their rivals and invaded neighbouring states, once-favoured African leaders have become targets of virulent campaigns by human rights organisations with international followings. DfID had promised the British electorate, and the rest of the G8, that it would massively increase aid. But once you subtracted oil- and mineral-rich African states that didn't need foreign aid, then removed those which were undoubtedly dirt-poor but whose leaderships were considered beyond the pale, the list of governments meeting the criteria for partnership became embarrassingly short.

There is also a manpower issue. While the British Treasury had promised to raise DfID's budget, it expected the ministry to cut staff numbers just like every other government department. With fewer staff available to disburse more funds, DfID is understandably keen on moving wherever possible from time-consuming project aid to direct budgetary support. But that requires confidence in the government concerned. 'They are desperately pushed by the need to disburse,' says Edward Clay. 'It's supply-side pressure. Most departments have the Treasury breathing down their neck to spend less. DfID is unique in that it is required to spend more, and farther away from scrutiny than any other department.'

Last, but not least, comes the China factor. In the last decade, resource-hungry China has been making sweeping inroads on the African continent. On the hunt for the oil, timber, copper and other resources needed by its expanding economy, Beijing offers in exchange cheap funding without the moralising lectures and conditionalities of Western donors. Just as Western governments thought they had put their colonial guilt firmly behind them and established a post-Cold War consensus on what was needed in Africa, a giant player dangling no-strings-attached funding – on governance matters, at least – enters the game. Yet another argument, in the eyes of development ministries suddenly facing their own irrelevance, for erring on the side of leniency.

Small wonder, given these various factors, that DfID in 2004 had little appetite for the antics of a high commissioner who appeared to have launched a personal crusade against government venality in a key African ally. 'They found it an embarrassing obstacle, because it got in the way of their plans to spend more,' says Clay. 'They found it unpalatable to have an ambassador who had a high-profile role on it and was not going to pipe down.'

By the time Clay arrived in Kenya he already nursed reservations about DfID's role. He was beginning to suspect that while the department dutifully recited the 'good governance' mantra, it essentially regarded the fight against corruption as an inconvenience. 'They've said for years, "Good governance is at the centre of development," and it's been easy because they thought their bluff would never be called.' Now he couldn't help registering that the amounts involved in the dodgy Anglo Leasing deals were roughly equivalent to the sums the Kenyans received each year in aid. Was international generosity merely encouraging those in power to feel they could help themselves to equivalent amounts?

A principle, he felt, was at stake: the link between voter and government in a young democracy. Critics of international aid often claim it all ends up in Swiss bank accounts, a charge development officials easily swat away, pointing at the accountants and consultants who police spending. The argument should be a different one: not that the aid is itself stolen, but that donors make it possible, via that aid, for governments to dip their hands elsewhere in the budget while still delivering basic services, thereby escaping the electorate's wrath. Accountability moves offshore, thanks to aid's fungibility.

Surveys in Kenya showed that the one area in which NARC consistently won top marks from otherwise disillusioned voters was its free primary education programme. In rural areas, families who had rationed attendance to an eldest boy could now send all their children to school. In actual fact, DfID funding made it possible for the government to keep class sizes down and make schoolbooks available. Yet, fretted Clay, NARC would get the credit come the elections, and might win another term on the back of DfID's input. If that happened, British taxpayers would effectively have shored up a corrupt regime, not only shielding it from the ire of its voters but buffing its image. Was that really the role they should be playing?

The identity of the minister for education rubbed salt into that wound. Finance minister under Moi, George Saitoti was a man long suspected by the donors of involvement in the Goldenberg scandal. Yet London yearly bestowed a £50-million blessing, and with it international credibility, on Saitoti's head. 'What the Kenyan public see is a minister of education who has been alleged to be a central character in Goldenberg, being chummied and rewarded by the British, who regard him as the most important single partner in their development work. The message that conveys should make us uncomfortable.'

During Clay's tenure, the topic of shifting from project to budgetary support in Kenya cropped up with a regularity that left little doubt where DfID's inclinations lay. The department was champing at the bit, desperate to make the move. 'Unfortunately, they've had to confront the evidence that they can't justify doing it, because the "fiduciary risk" was too high. What they have tended to do is to shoot the messenger.' And the tension between DfID's need to spend and its prospective customer's soiled track record could only, Clay recognised, get keener with time, increasing the temptation to turn a blind eye to government theft. 'DfID have got themselves into a position where they talk about not wanting to waste "their investment" in Kenya's education. It isn't really an investment, of course, it's an expenditure. But when people talk about investment in that way, they start thinking about defending it, and the more they invest, the more they will be inclined to say, "We mustn't pull out now."' Kenya's education system represented an expanding 'sunk cost' for DfID, its abandonment less acceptable with every passing year.

In the high commissioner's view, Britain's integrity was on the line. He, for one, had taken on board Whitehall's briefings on a new line on graft. 'Governments have been telling us diplomats for the last ten years: "You have to take up the moral high ground." My argument was, I'm in Kenya, we have the evidence, this is something we can do something about. For God's sake let's do it if we mean what we say.'

It was not a question, in Clay's view, of cutting aid. More subtle weapons were available, as his own high commission and the American embassy would later demonstrate when they started refusing visas to Kenyan politicians who saw shopping expeditions to Harrods and weekends in Manhattan as a right. The public humiliation cut the likes of Chris Murungaru to the quick. It was a question of engaging forcefully with the government and making clear, loudly and tactlessly if necessary, that Britain expected more than lip service from its African partners in the war on graft. It expected, at the very least, not to have to discuss development with ministers widely suspected of having had their fingers in the pie. 'We should be less niminy-piminy and say what we think.' If he felt he owed it to British taxpayers, Clay also felt beholden to the Kenyans he met. Given the dangers faced by those trying to make government accountable – arrest, raids by the security services, legal writs – keeping quiet undermined those attempting to reshape their own societies, the very individuals donors claimed to want to encourage.

Most Kenyans blithely assumed that if the British high commissioner said something, it represented British policy, a thought-through position running from one end of government to the other, like the lettering in a stick of Brighton rock. Not so. Jarring with the upbeat mood music in the run-up to Gleneagles, Clay's call for a tough line on Kenyan graft set him at odds with DfID, his own bosses and a Downing Street operation famous for its control freakery. Tony Blair's administration was not one in which individual departments were supposed to break free from their moorings. Unbeknownst to Clay, Vomit One had already triggered an exasperated outburst in the unit drawing up the Africa Commission report, preparing to tout the line that a reformed continent was perfectly placed to benefit from increased aid. 'Who is this guy?' spluttered an appalled senior Treasury official involved in its drafting. 'Why aren't they reining him in?' No one wanted a repeat. The issue was not 'scaleable', Clay was told by his superiors, meaning events on his patch could not be used as fodder for overarching government policy. 'We were running up to the Year of Africa and they didn't want someone pissing on their parade.'

The British New Year's Honours List for 2005 contained a knighthood for Clay, the ultimate compliment from an appreciative government. But he was becoming a little too pungent for his peers, both in London and in Kenya. Attending a conference on corruption in Nairobi a few months earlier, an event which attracted the usual crowd of badge-wearing NGO representatives, diplomatic envoys and Kenyan government ministers, I was nudged in the ribs by a journalist friend. He pointed to where Clay sat towards the back of the hall, looking determinedly cheerful but very much on his own. Arriving diplomats greeted him, but chose to sit elsewhere. It was as though he was possessed of an invisible aura repelling incoming bodies. 'Look. No one wants to be associated with him now. He's become radioactive.'

He gritted his teeth and ploughed on, ignoring London's heavy hints. 'I was doing the right job, with excellent material. I was buggered if I was going to be co-produced from abroad.' On 2 February 2005, Sir Edward chose a press awards ceremony funded by Britain as the occasion for Vomit Two. The ceremony at the Hotel Intercontinental, meant to recognise Kenya's investigative reporting, became an exercise in irony. Mutterings of embezzlement circulated the tables as it emerged that two competing lists of prize-winners existed. Vice president Moody Awori had graced the event with his presence, and the organisers must have sensed what was coming, for they did all they could to curtail Sir Edward's performance, packing the programme with extra speakers and knocking on the podium as he delivered his speech to try to bring it to a close. If Vomit One had marked a general sounding of the alarm, Vomit Two, Clay's second salvo in six months, was far more specific. His team had done their homework.

The 'unpleasing substance' he had previously cited remained firmly stuck to the shoes of both Kenyans and the donors, Sir Edward told his audience. He had therefore handed to the Kenyan authorities details of twenty suspect procurement deals. What all these arrangements shared was the customer – usually 'the good old OP' (Office of the President) – and a certain kind of businessman. 'At the back of all these questionable deals is a type of man whose companies travel under many colours and names, but who goes on, apparently forever. He is a man for all seasons of governments. He can find receptive palms in every political party, but you should count your fingers after you shake his hand. He finds changes of government no more alarming than changing his shirt. He has plenty of them; unlike poor Kenyans who lose their shirts.' As for apologising for Vomit One, he declared, in full Edith Piaf mode, that he regretted only three things about it: not having spoken earlier, underestimating the scale of the looting, and the moderation of his language. He concluded with a parody of another famous children's poem, this time T.S. Eliot's 'Macavity: The Mystery Cat'.

> 'He likes to be in transit, and he's partial to hotels,
> He has a place in Manchester, he's fond of the Seychelles.
> So when the nation's revenue's in European banks,
> Or you need a team of tractors, but acquire a troop of tanks,
> Or the nation's full of caviar, but hasn't any bread,
> Or you want a road for Christmas, but a frigate comes instead,
> You can look behind the scenery or stare up in the air,
> But the ministers will tell you that Macavity's not there.'

It was another Clay barnstormer. Sir Edward's manner of expressing himself might have been as florid as ever, but he had clearly grasped the fundamentals of the Anglo Leasing deals: that they formed a continuum stretching across the 2002 elections; that Moi's elite had passed its appetites, habits and – most importantly – sleazy business contacts smoothly on to Kibaki's set.

The government frothed. A furious minister accused Sir Edward of being an 'incorrigible liar' and drunk. Yet despite its more weighty content, Vomit Two would have far less impact than Vomit One. In the normal state of affairs, the Kenyan media would have run with the story for weeks. But an upcoming event was about to seize their attention, banishing any other event to relative obscurity.

The months that followed John's non-transfer were not proving happy. The mysterious paybacks continued, and Anura Perera stepped up his efforts to secure a tête-à-tête John was determined would never take place. Even as he received ever more detailed information about the Anglo Leasing contracts – facts he dutifully passed on to his head of state – Kibaki delivered a series of speeches protesting that he could not take action against alleged government graft 'without evidence'. The hypocrisy made John wince. What further proof could Kibaki possibly need?

With the passage of time, the Mount Kenya Mafia grew ever more careless, taking John's acquiescence as read. Its members moved with disconcerting ease from pretending to know nothing about Anglo Leasing to coolly presenting it as part of a 'resource mobilisation' effort being pulled together by internal security minister Chris Murungaru and Alfred Getonga for Kibaki's DP party ahead of elections due in 2007.

So here, then, was the supposed justification for a million-dollar con being perpetrated on the Kenyan taxpayer. Elections cost money, as any political party knows. Moi's twenty-four years at the helm had given KANU every opportunity to build up a vast war chest, the argument went. If NARC was to stand any chance of blocking KANU's return in the polls, or pricking Luo leader Raila Odinga's ballooning ambitions it needed cash. It was unsavoury work, but someone had to do it. 'If your pig gets stuck in the mud you have to jump in and extract it, even if that means getting dirty,' Kiraitu Murungi told John.

John knew enough about his colleagues' appetites not to give their explanation any credence. 'With these kind of arrangements, only 30 per cent ever goes on political finance and 70 per cent goes on personal spending. It wasn't about electoral funding. It was about BMWs and mistresses.' NARC's 2002 election win, after all, had been built on authentic popular support, not bought votes and war chests. And what appalled him were the terms in which the argument was couched. 'What staggered me was that the justification was put in ethnic terms. It was: "We need to protect ourselves from the Luos." "Who is 'we'?" I'd say. It was always implicit. It meant "We Kikuyu, Meru and Embu."' Since independence, Kenya's various elites had used the clarion call of ethnic solidarity – 'We're doing this for you, our brothers!' – to camouflage grotesque levels of personal enrichment. Once again, tribal rivalry was being used as the cover for theft. By late October, he had calculated that the suspect contracts could amount to over $1 billion, for behind the eighteen Anglo Leasing cases lay other, even more secret and even murkier projects. The scam just kept growing. 'I'm worried that we will have another Goldenberg scandal before the elections,' confided a clearly uneasy Kiraitu Murungi. Part of John watched himself, horror-struck, marvelling, as he joked with the justice minister – that macho joshing again – about the 'Goldenberg scandal of 2006', speculating about whether they would appear as suspects or witnesses in the inquiry.

In every individual's life there are areas that remain permanently off limits to the biographer, defying explanation. Tracking John's itinerary, there's something mystifying about the sheer time it took him to recognise the obvious. The dossier he eventually produced can read like a log of a year-long refusal to face the truth. How many times did John Githongo, a man of no mean intelligence, need to be told that his closest colleagues had hatched Anglo Leasing on the pretext of election fundraising before he believed it? Time after time he recorded conversations in which he was told exactly that in his little black Moleskine diaries, concluding in despair: 'I must now prepare to leave the administration.' Two weeks later, an almost identical exchange would be recorded, with a touch more emphasis: 'I really MUST make arrangements to leave.'

His family's welfare preyed constantly on his mind. But the problem went further than that. John was always ready to admit that procrastination, which follows on from the need to control events as night follows day, was one of his character flaws. 'I haven't even started procrastinating yet. I'm going to start tomorrow,' he used to joke, poking fun at his own failing. His girlfriend dubbed the reflex, a tendency to flagellate every decision to the point of near-extinction, 'analysis paralysis'. Friends saw this Hamlet-like irresolution as the expression of a perfectionism that had characterised John since school. If the unexamined life is not worth living, the over-examined life can become a terrible burden. 'If you aren't free to make mistakes, you aren't free to act. No mistakes are tolerable to him, and that accounts for the inaction,' says Mwalimu Mati.

But that was not the only reason for the foot-dragging. John's slow mental churning, so baffling in retrospect to the clear-eyed outsider, was also a form of mourning. 'It sounds almost comical, like I'm describing someone in a relationship. But I honestly think John's heart was dented,' says former colleague Lisa Karanja. Like an accident victim sitting dazed in Accident & Emergency, John needed to go through the necessary stages of disbelief and denial before he could digest his loss. The time it took – not months, but years – was a testament to the depth of his belief, in that bright post-election dawn of 2003, that NARC could break with Moi's grubby ways. For John, who like so many Kenyans had believed he was taking part in an ethical rebirth, this really was a grieving process, and it could not be hurried. The oedipal aspect of the ordeal made it all the more painful. A young man who had enjoyed a filial relationship with the president was being forced to realise that Kibaki had feet of clay, just as he had once registered with his own father. When acceptance finally set in, affectionate belief would be replaced by cold fury.

By October, John was setting his affairs in order, still not entirely certain how far he was willing to go. Whatever happened to him, he did not want his staff to bear the brunt. Quietly he downsized his unit, sending civil servants back to the departments from which he had poached them.

There was to be one last betrayal, in the course of one of the sparring conversations that had become a surreal feature of his working life, during which deadly-serious matters were alluded to between guffaws of phoney laughter. Following a 4 November meeting to discuss NARC's governance strategy, Justice Aaron Ringera, head of the Kenya Anti-Corruption Commission, stayed behind for a private chat. Fraught, craving support, John ruefully confessed to the colleague whose career he had done so much to further that he had realised State House only wanted him to go through the motions of his job: 'Mine is the shock of a personal realisation.'

Ringera nodded in agreement. 'So you stay there, you are a little wiser and you know that you are there!' And then this refined man of the law, John's trusted friend, made crudely explicit the sheer horror of his predicament. 'You can't, in fact, afford to make any move. That's when you will really be killed.'

So, joked a dismayed John – attempting, as ever, to clothe the abnormal in palatable language – the message from his superiors was: 'That's far enough.'

Ringera nodded, and added a diabolical twist: 'If you wanted to resign and go today, that's when they would kill you.' It was the classic predicament of those who climb onto the merry-go-round of power, only to find themselves whizzing around so fast they cannot jump off.

The two chortled cynically together as John sketched aloud a twee scenario in which the hulking anti-corruption chief drove around in his 'little car', doing meaningless errands while sending out reassuring messages on the government's behalf. If Ringera was right, John was a prisoner in his job, damned if he tried to do it, damned if he resigned.

The conversation was one of the bleakest moments of John's life. The man who should have been his greatest ally had just passed on a death threat. 'I felt lonely, very lonely. I realised at that moment that I had no more allies within government. In just one and a half years, it had come to this.'

On 12 December, Kenya marked the anniversary of its independence. Jamhuri Day was when the presidency handed out its yearly honours, and to John's surprise he was made Chief of the Burning Spear, First Class. The message was none too subtle: 'Keep quiet and you will be amply rewarded,' his bosses were telling him. Two days later, he was called to the ministry of finance to meet donor representatives. He was being used to lie to Kenya's backers, a pet monkey performing tricks to reassure the regime's critics. It was a miserable Christmas and a grim New Year.

His last briefing with the president was on 10 January 2005. Shortly before the meeting, Kiraitu had been quoted in the national press stating that because the money involved in the first two contracts had been repaid, Anglo Leasing could be defined as 'the scandal that never was'. No money was lost, nothing had actually happened. Not so, John told Kibaki. Running through the eighteen contracts, he doggedly drew the outlines of the scandal in as much detail as he could. The usually torpid president, a man often described as too lazy to lose his temper, suddenly grew animated. For a brief moment, John's heart lifted: had Kibaki finally become outraged at the scam's depth and breadth? No, he realised. He was incensed by the uppity young whippersnapper who insisted on making his life so difficult, harping on relentlessly on this discomforting topic. 'We were both very angry. He was furious, we couldn't look each other in the eye. Our relationship had collapsed.'

If the series of paybacks from phantom companies had briefly suggested John was cramping the fraudsters' activities, even that victory was slipping away. Joseph Kinyua, permanent secretary at the finance ministry, informed John he was coming under tremendous pressure from David Mwiraria to pay out on some of the Anglo Leasing contracts. Kinyua said he hoped to use auditor general Evan Mwai's report as a delaying tactic, but Mwai informed John that Francis Muthaura, head of the civil service, had been telephoning to insist pending bills were paid, even if the audit was incomplete. John tried to stiffen each man's backbone, encouraging them to hold firm. But the money-making machine brooked no delay.

As it gathered momentum, John's colleagues seemed to lose all reserve. There was a palpable change in tone. 'You are burning down the house,' they told him, meaning the House of Mumbi. 'You cannot burn down the house to kill a rabbit.' No attempts, now, to hide what was going on. After a cabinet meeting at State House on 14 January, Kiraitu walked into John's office, pointed at him and accused him of undermining the party. According to the minister, the forthcoming party elections were expected to cost 200 million shillings, and those funds would come from Anglo Leasing. The devastatingly frank conversation was repeated a few days later, with both Kiraitu and Murungaru openly telling John the suspect contracts were all covers for party financing. 'They had bared their souls to me.'

There is a famous Sherlock Holmes story – 'Silver Blaze' – which features a dog that fails to bark. A precious racehorse has been stolen, but the clue exposing the guilty man is not a footprint or blood smear, but a non-event: the guard dog fails to raise the alarm because the criminal is, in fact, his master. With Anglo Leasing, just as in 'Silver Blaze', an absence, not a presence, was the giveaway. From the start, John's priority had been to establish where the president stood on the matter, because in that answer, he knew, lay a nation's success or failure. No paper trail led from the bogus deals to Kibaki, but that was hardly surprising, given most African leaders' preference for keeping instructions verbal. The ministers' brazen behaviour told John all he needed to know. The truth lay in what they *didn't* do, not what they *did*. Had Murungaru and Kiraitu been briefing him on the sly, whispering their insights to him in his office, looking over their shoulders, it would have indicated the existence of factions and cabals in State House outside Kibaki's control. But no, by January 2005 both men were strolling into his office and, without any sign of embarrassment or anxiety, confessing all. Only Kibaki's blessing could explain such blithe self-confidence. 'The fact that these guys were openly, without concern, admitting to these things, knowing I was reporting back to the president – while Kibaki was making public statements saying "Where is the evidence?": but I'd *given* him the evidence – said it all.'

In conversations with diplomats, with the head of the KNCHR or the director of public prosecutions Philip Murgor, John rigidly held the party line. 'Maina Kiai and I were telling John, "We're seeing all the signs, we're on our own,"' says Murgor. 'He kept saying, "I can assure you, I was just with the president and he is behind us." John remained loyal till the very end.' But John could no longer hide the truth from himself. 'Ultimately, it became clear. I was investigating the president.'

Enough: the decision was made. One of his last errands was a visit to Archbishop Ndingi Mwana a'Nzeki, head of Kenya's Catholic Church. Strong-minded clergymen have played a pivotal role in Kenyan history, nudging reluctant governments along paths they hesitated to take. The archbishop, John knew, had recently had a heated encounter with Kibaki, delivering a series of criticisms on his bishops' behalf so scathing that the president – supposedly a devout Catholic – had stormed out. It made tactical sense to see him. But John also wanted to consult the archbishop as a practising Christian. He wanted the Church's backing, implicit if not explicit. 'Things are more complicated than I expected. And they go much higher,' he said. 'Yes, that's what we heard,' the archbishop replied, equally elliptically.

At the eye of a hurricane lies a still, calm place where release finally becomes possible. Back at his villa in Lavington, John larked about with girlfriend Mary, taking pictures of the two of them in the house and gardens. He had barely had time to settle in, and now he was aware he might not get to enjoy such lavish premises for many years to come. His family were kept in the dark: too risky to tell them. But Mugo twigged immediately when he drove round to see John and came upon him burning documents. 'That's when I knew he was off.' Having long fretted over John's government liaison, his younger brother was quietly delighted. Thank God he was pulling out before becoming any more tarnished.

John's trip would take him first to the Swiss resort of Davos, then to Oslo and London, where he and Justice Aaron Ringera were supposed to be tracking Kroll's progress on Goldenberg. In the flurry, no one noticed that this experienced traveller, the kind of man one would normally expect to see pulling a single roll-on case – seemed to be carrying an unusually large amount of luggage. In London John and Ringera, the man who had disappointed him above all others, held a last tête-à-tête, walking around Horse Guards Parade for two chilly hours. As they crunched across the gravel, John prepared Ringera for what was coming: his resignation and exile. Leaving Kenya would ensure the spotlight of government scrutiny shifted off his family and followed him abroad, leaving those he loved unscathed. And Ringera, a handsome, white-haired veteran who resembled everyone's favourite uncle, revealed the steel beneath the velvet. Kenyan intelligence would 'put something in your tea' if John went public with what he knew, he said.

'We made a sort of deal,' says John. 'I wouldn't attack them and they wouldn't attack me. And Ringera said he would take that message back to the *kiama* – he meant himself, Muthaura, Murungaru, Mwiraria and Kiraitu – and a truce would be called.' The judge's choice of terminology gave John a jolt. In its own small way, it validated his decision to leave. '*Kiama*' is not a Kiswahili word. It means 'council of elders' in Gikuyu and Meru, and refers to the gatherings which traditionally decided community matters in pre-colonial times. Ringera had just confirmed what every jaundiced non-Kikuyu Kenyan had suspected since 2002: power in modern Kenya rested in the hands of a narrow ethnic clique. 'I knew there was a mafia. Every mafia has a godfather, and the godfather was Kibaki, and that's why I left. But it was the first time I'd heard Ringera refer to it like that.' State House had been primed. As I toured my north London district looking for somewhere from which to fax John's resignation letter, the Kibaki administration was already braced for what was to come.

For two years, John's presence in government had served as a form of shorthand telling Kenya's foreign partners that despite a few blips, this was still an administration worth trusting. 'You just didn't get it, did you?' *Nation* editor Joseph Odindo would later tease John. 'You were meant to be a fig leaf.' The fig leaf had now been whipped away. How, the Mount Kenya Mafia uneasily asked itself, would the donor community which had taken such a shine to the charming anti-corruption chief react?

Pretty limply, was the answer. Ambassador William Bellamy announced that the United States was suspending the annual $2.5 million it gave to John's Office of Governance and Ethics, the KACC and the ministry of justice. 'It makes no sense, obviously, to partner with a government whose commitment to improved governance is purely rhetorical,' he said. It sounded good as a headline, but in the same speech Bellamy revealed that the US allocated $175 million a year to Kenya: Washington had frozen just one seventieth of its yearly aid. Eight key embassies issued a joint statement describing John's departure as 'an extremely serious challenge' to the credibility of the government's anti-corruption policy, and the EU warned it might reconsider aid if the government did not show 'seriousness and a sense of urgency' in addressing graft. Given the many months in which top Kenyan officials had demonstrated their total lack of seriousness towards any activity other than stuffing their pockets, the statement seemed disingenuous in the extreme.

Two months after John's departure, the donors got their chance to show the government exactly what they thought of its behaviour. Held at a sprawling hotel complex on Nairobi's outskirts, the Consultative Group meeting on 11–12 April 2005 was one of a series of get-togethers held sporadically between donors and government, occasions at which the administration's performance was evaluated and fresh pledges of funds made, with civil society invited along to lend its voice. With its anti-corruption supremo just turned international fugitive, the government was braced for attack. A reshuffle in which Kibaki had demoted internal security minister Chris Murungaru to transport minister looked unlikely, outsiders speculated, to placate the donors. 'They were running scared, on the ropes,' recalls Clay. In an attempt to forestall its critics, the Kenyan government had submitted a two-year action plan for tackling sleaze. 'We'd spent the previous weekend crawling over it,' says Clay. 'It was mostly things they'd said they would do in other forums, all good things but dependent on the political umph behind them. We had decided to offer some amendments, we were going to go hard on that.'

The work proved in vain, thanks to the efforts of one man. As World Bank country director, Makhtar Diop had automatically been chosen to chair the get-together. Forceful chairmen can shape meetings to go the way they please. Privy to all the donor discussions ahead of the CG meeting, Diop was well aware of the intensity of concern, yet after the finance minister David Mwiraria had made his presentation, he did his best to close down all mention of corruption, delaying discussion of the action plan until the evening, when everyone would be itching to get home. 'Makhtar had decided to shrink the time available and suppress the points we'd prepared. He did his best to subvert that meeting,' recalls Clay. The US embassy's first secretary was forced to physically grab the microphone to insist his ambassador had a say. 'I was either going to interrupt or walk out,' remembers William Bellamy. 'The meeting had turned into a series of lengthy lectures from government spokesmen, with no chance for the donors to respond.' Gladwell Otieno, TI-Kenya's new director, was left fuming after trying to make a contribution. 'He wouldn't let me speak!'

The outgoing World Bank director's behaviour appalled all but the government delegation. A chance to signal that the international community would not swallow one outrageous scam after another had been missed. It's striking how angry those present that day still feel, years later, at Diop's act of sabotage. Had his overly cosy relationship with his landlord – yet to be soured at this stage by Lucy Kibaki's behaviour – played a role? What seems clear is that he had fallen prey to a version of Stockholm syndrome. Over-identifying with his client, he had lost sight of the fact that his role was not to argue the Kenyan government's cause. 'Makhtar wanted to leave on a high note and to be thought a hell of a guy, with plaudits from the Kenyans and signed pictures. Well, who doesn't?' says Clay. 'His behaviour that day was an absolute scandal. He really did the donors a bad turn, revealing to the Kenyans that there were more important things than governance, or morality, and one of those things was his own vanity. The action plan was approved with only minor changes, and flowed into the sand soon after that.'

As far as the regime was concerned, it was business as usual. A man whose presence had been regarded by all as a litmus test of its good intentions had resigned in a way that made clear something was very, very wrong at the heart of government, and – after the briefest of wobbles – State House had got away with it. Few were surprised when four days after the donors' meeting, the militant Gladwell Otieno was forced to resign from TI-Kenya by the same *wazee* who had once recruited John. A second high-profile anti-corruption campaigner was gone, from a none-too-crowded field. The regime had received the signal it wanted from the donors, however grudging, and saw no reason to hold back.

Clay's last few months in Kenya would be a difficult time. His health was playing up – stress-related, perhaps – and as he underwent a series of medical scans he sensed a falling-away by his colleagues. Relations between DfID and the FCO in Kenya had become 'awful', recalls one junior officer: 'One side was saying it was white, the other was saying it was black.' When Clay talks about this period now, his voice becomes light and whispery with remembered strain, so thin the tape recorder barely picks it up.

In June 2005 the Clays headed back to their home on the chalky foothills below Epsom's racecourse, a semi-detached house on a middle-class estate so modest any retired African high commissioner would regard it as evidence of a wasted career. Before leaving, he tried in vain to persuade local DfID staff to meet and discuss a common approach on corruption. 'From July 2004, no one in the FCO in London and no one in DfID in Nairobi was willing to debate the issues. DfID staff always had something better to do. That demoralised me to a huge extent. It was clear to me that when I left there would be a marked rowing back on the past.'

13

In Exile

'. . . Now whether it be
Bestial oblivion, or some craven scruple
Of thinking too precisely on the event
. . . I do not know
Why yet I live to say "This thing's to do." '

WILLIAM SHAKESPEARE, *Hamlet*

As he began his exile in February 2005, John could be under no illusions as to the life-and-death nature of the choices before him. A few days before his resignation, he'd been presented with the most chilling lesson imaginable in the dangers of challenging the powers that be.

Georgia's prime minister, Zhurab Zhvania, had been the undisputed star of an anti-corruption conference staged at Nairobi's Safari Park Hotel the previous October. Zhvania had played a key role in the Rose Revolution that had toppled president Eduard Shevardnadze after his clumsy attempt to rig the 2003 Georgian elections. As prime minister, Zhvania had dismissed most of Georgia's army generals, slashed the size of his own office and scrapped the traffic police unit, a key source of sleaze. His no-nonsense advice to delegates had been electrifying. When a regime has been toppled it is vital to strike fast, before the power blocs that perpetuated graft had the chance to regain their balance, he had told the audience. 'Use your chance. You will not have another.' That message had contained a certain personal irony at the time for John, horribly conscious that NARC had almost certainly already missed its chance. Sharing a podium, John and the Georgian prime minister had hit it off. Like John, Zhvania had been subjected to a smear campaign at home, accused of being a foreigner and a homosexual. John had been touched by the fact that Zhvania had brought his young son to Kenya with him, explaining that his job allowed him little time with his family, and the trip was an opportunity for some precious paternal bonding.

That young son was in mourning now. Zhvania's bodyguards had found his father's lifeless body in an armchair in the Tbilisi apartment of an associate, also dead. The official explanation was carbon monoxide poisoning, caused by a faulty gas heater. A subsequent FBI investigation would find that levels of carbon monoxide in the flat were not high enough to kill, leading Zhvania's family to conclude that the two men had been murdered elsewhere, their bodies brought to the flat and arranged to make their deaths appear a banal domestic accident. Dead at forty-one, Zhvania had probably paid with his life for the very take-no-prisoners approach that had so impressed the Nairobi conference.

It was a dreadful corroboration of Justice Ringera's warning, but other cautionary tales urged John in the opposite direction – to take action. He knew former officials who had served under Moi, watched procurement scandals break on their watch, and retired. 'They left government, kept their mouths shut and went into private practice. And look what happened. After that came queue voting, the Ouko murder, ethnic clashes, and Goldenberg. Letting the system repair itself doesn't work. I'm left thinking: what would have happened if they had taken a stand?'

By March 2005 John had established his base in Oxford, at the city's newest and most international graduate college. Founded in 1950 – absurdly recently by Oxford's standards – St Antony's neither looked nor behaved like a traditional Oxbridge college. While most colleges boast a small contingent of tenured fellows who have the right to live out their days on site, St Antony's had embraced the notion of 'senior associate members', a title conferred on foreign visitors temporarily setting aside their careers to don academic robes. With sixty senior associate members on staff at any one time, the college prided itself on being more open to the outside world, more invigorating, than its more ancient rivals.

Back in Kenya, many of John's friends had groaned when his new country of residence became clear. 'He should have gone anywhere else: Norway, Sweden, Canada, anywhere, just so long as it wasn't Britain,' one told me. The choice played into the hands of John's enemies, who could cite it as final proof that Githongo had been a British snitch, working to destroy the NARC administration from within all along. The situation was not helped by the fact that St Antony's College, whose first warden was World War II soldier and adventurer William Deakin, included several crusty former members of Britain's intelligence on its staff and was popularly rumoured to be a recruiting ground for agents. But from John's perspective, Britain was always the obvious destination. It was where he was born – which meant he didn't need to apply for political asylum – and where he had gone to university. It felt like a second home.

Detractors in Nairobi whispered that he was living the life of Riley, richly rewarded by his British paymasters for his perfidy. So convinced were many by these claims that the St Antony's switchboard got the odd call from angry Kenyans demanding to know how much the college was being paid by MI6 to lodge this known British spy. In fact, while John's title of 'senior associate member' sounded impressive, his living arrangements were as bleakly functional as those of any postgraduate student. St Antony's provided lodgings – a one-bedroom apartment on the top floor of a house on the Woodstock Road – and space was cleared in the study of a college-owned Victorian house where a computer and filing cabinet were made available. Sharing this room with a succession of scholars, John would sit tapping at his keyboard, fuelling himself with Red Bull and cups of tea brewed from the kettle in the corner, as the driving rhythms of 'Green Onions', that undergraduate perennial, wafted down from the rooms above.

At the age of forty, John had left the world of bodyguards and cocktail receptions, returning to the Spartan life of the student bachelor, doing his own washing, fixing sandwiches in a kitchenette, measuring out his life in instant coffee spoons. He was now an ordinary citizen, and one, ironically, whose credentials won him not fawning respect but suspicion. When, trying to open an account at the local branch of Barclays, he mentioned that he had worked in the office of the Kenyan president, the process suddenly hit an invisible buffer. 'I had to get a special letter from the college authorities. Barclays realised I was the kind of PEP – Politically Exposed Person, to use the lingo of the corruption world – who could be depositing a large amount of illicitly-acquired money.' For someone who had expended so much energy trying to police Kenya's PEPs, the exchange was grimly amusing.

John was soon made aware that his watchers had followed him from London. Coming out of his study one day, he noticed a smartly dressed black woman outside the house, mobile phone clamped to ear, perched in a position from which she could survey his comings and goings. She made no attempt at concealment. In fact she seemed to be deliberately making her presence felt. While some of his 'shadows' in London had borne the hallmark of Kenyan intelligence, this, to John's eyes, looked more like the employee of a private firm of detectives, hired perhaps by one of the Kenyan businessmen whose profits he had jeopardised. 'We're still watching you,' they were telling him.

He was always careful never to speak long on his mobile, aware that private sleuths, unlikely to suffer from the Mount Kenya Mafia's technological backwardness, only needed a tracking device set up across the road or in a parked car to tune in. The wisdom of that policy was highlighted when he received an unnerving call from his mobile phone provider:

'We're calling to confirm you received the fax we just sent you, Mr Githongo.'

'What fax is that?'

'Didn't you just ask us to fax you itemised copies of your recent bills?'

'I've no idea what you're talking about.'

'But we just got a call from you asking for exactly that . . .'

The impostors were so sophisticated they had sailed smoothly through the mobile phone company's security checks, obtaining in the process several months' worth of itemised bills, a priceless guide as to how John got his information and who he spoke to. He changed his passwords, but it was a valuable lesson not to let his guard slip.

The role played by the burly college porter, an Oxbridge institution, came as a welcome revelation. Often recruited from the ranks of retired policemen, college porters act as a cross between receptionists, postmen and bouncers. As any student staggering back drunk in the early hours learns, nothing delights them more than an opportunity to put their old skills to the test. At St Antony's these professional talents were particularly finely tuned, given the college's practice of welcoming controversial guests. Before John, there had been former Malaysian deputy prime minister Anwar Ibrahim, who like him had been smeared with the homosexual brush after accusing his prime minister of corruption. Another senior associate, a Colombian, had been gunned down after returning home to campaign in his country's presidential campaign. Notified by the college authorities that their new African student was one of these 'special' guests, the porters closed around John like bulldogs. 'These chaps have really formed a ring of steel around me. I've learned new lessons about professionalism, dignity, a sense of power and tradition from them.'

He appreciated their understated, lugubrious humour. Soon after John's arrival, the head porter had called to report that a suspicious-looking package had arrived with his name on it. When John asked, only half-jokingly, whether it was ticking, the head porter solemnly replied: 'I will immediately dispatch my most dispensable porter to give it a firm kick and find out, sir.' When John popped into the Porter's Lodge to mention that a visitor from Nairobi – someone he did not want to meet – might come asking for him, the news was met with grim relish. 'The porters love that kind of challenge. You can see their eyes light up. "Is that so?" they'll say, and later they'll mention casually: "Oh, we sent them on their way." St Antony's is a nasty place to try and get into if you're not invited.' If John didn't want to see you, you might as well not turn up. And if you phoned the college switchboard asking for him and didn't give the correct extension, your call would not be put through.

But John still felt vulnerable enough to call on the services of the police. Officers from Special Branch came to inspect his premises, providing an emergency number to be called if ever he felt in physical danger. The system was later upgraded, with alarm buttons installed in both study and bedroom, and John given a mobile device to carry around, which he promptly forgot. Having been in a constant state of alert for two years, he could no longer summon the psychological tautness of old, as he discovered on the very few occasions when he tried taping conversations. 'Since leaving Kenya, I've never been able to use a wire with any success. I've lost my competence.'

The only other form of self-protection he could muster was his old stalwart: information. Knowing what was happening back in Kenya, perhaps better than many of the residents of State House, was a weapon of a sort. The network of informants he had established when in office had not evaporated with his departure. If some assumed that with resignation came irrelevance, many continued feeding him snippets. Now, no one expected to be paid. They acted out of a sense of civic duty, the fury of frustrated zeal, or simply took a punt on a distant future: 'One day you will be a Big Man and you will look after me,' some texted him, invoking the very patronage principle John regarded as holding the key to Kenya's psychosis. Electric cables trailed across his small flat, an adaptor was jammed into every power socket, keeping the eleven mobiles on which this information-gathering operation relied constantly charged.

There was an East African Studies programme to be pulled together, in collaboration with St Antony's existing experts, who included David Anderson, a specialist on Kenya's Mau Mau, and Paul Collier, who had spent thirty years researching the causes of African poverty. While doing that, John fended off an onslaught of interview requests from journalists, foreign diplomats, NGO workers and

Kenyan acquaintances. Many were genuinely concerned to see how he was doing. Others were simply curious, keen to be able to boast that they had met the former anti-corruption chief in his Oxford lair. Some, he knew, would have agreed to serve as the eyes and ears of the Kenyan government. The only entity that showed no intense interest in his presence – what an irony, given the accusations being levelled back home – was the British government.

For the most part he ignored the appeals. His move to Oxford would see the return of an idiosyncrasy long familiar to his friends, now exaggerated to cope with the sheer volume of invitations coming his way. Being Githongoed became an obsessive topic of rueful discussion amongst those who knew him. When the idea of a book was first mooted, I told John I would only press ahead if he approved. He gave the thumbs-up, but this, I learnt, did not mean preferential access. Every interview would feel like a triumph of perseverance, a small miracle pulled off in defiance of his constant urge to cancel our arrangements. I would be Githongoed over the mobile as I headed to the railway station; Githongoed on the train itself, hurtling towards Oxford; Githongoed on the very doorstep of his college study, finger on buzzer. I once spent a weekend staying with friends who lived a hundred metres from St Antony's being consistently Githongoed, returning to London three days later with nothing to show for my persistence. The only thing that made being stood up by John less painful was knowing I wasn't alone. We could have set up a club: the Jilted Friends of John Githongo.

The man, after all, was busy. For the very first time, all his Anglo Leasing evidence was gathered under one roof, and his focus now was on knitting together the material he had called in from his various *postes restantes*. It was not, he soon realised, going to be a quick job. Transcribing just one hour of taped conversation took at least four hours. And he had hours and hours of conversations to get through. He briefly considered taking on an assistant to help, but the problem was that while most of the conversations were in English, the quality of the recordings was so poor that it took an attuned pair of ears to make out what was being said. Anyone who has recorded a press conference, when the microphone is within inches of the speaker's lips, will know that the Hollywood scene in which the wired-up insider whips out his tape recorder and plays back the incriminating conversation, each word clear as a bell, is a Hollywood fantasy. Despite later breathless speculation as to how John had made his recordings – some would say the device was hidden in his spectacles, a pen, a watch – his equipment was rudimentary, and the results would have had any professional soundman tearing out his hair. He knew studios specialising in the clean-up of scratchy recordings existed, but they were far too expensive for him to use with all but a few key tapes. It made a huge difference if the transcriber recognised the various voices and knew the context of each discussion, which effectively meant that only John could do the job.

Even as he was carrying out these preparatory tasks, the equivalent of a surgeon laying out his scalpels, John received a call from Kiraitu, phoning while on a trip to Germany. His colleagues, said the justice minister, thanked him for keeping his counsel. His discretion had been noted. Given the task on which John was engaged, the conversation held a certain irony.

Working his way steadily through the material, John was aware that his perceptions of the recent past were shifting. In Nairobi, the adrenalin of unfolding events had kept him on his toes, but the excitement had obscured his vision. Swept along, he had recorded conversations without absorbing their full implications. 'I'd been too close, too close.' Spreading the events of the last few years before him like a map of his life, John now matched entries in the black notebooks with photocopied documents and the downloaded contents of his taped conversations. From what had seemed at the time a messy chaos, new outlines now emerged, like a submarine rising from impenetrable depths. Why, here, back in June 2004, was a diary entry recording a conversation in which Kibaki told him to stop his investigations, to back off. 'That was nine days before I was sacked. I came back, I wrote it down in my diary, but rereading it, I'm amazed. The words didn't really go in. I didn't digest it. There was so much else going on I put it to one side and carried on.' Players who had seemed peripheral emerged with fresh prominence: he could trace the series of red herrings he had repeatedly been served up by Francis Muthaura, head of the civil service, track the sinister manipulations of Kenyan intelligence. Like a carpet-maker crouched over a loom, John had allowed his eyes to become distracted by the criss-cross of colours, haring along after individual threads. It was only now, stepping back to survey the fabric in its entirety, that he could make out the overall design.

In July, very abruptly, he stopped. 'I suddenly hit a brick wall. I just couldn't bear to go on.' There were still hours of conversations to get through, but he had heard as much as he needed. Listless days followed one after another, but John was incapable of work, engulfed by lethargy. He had always been someone who embraced the night, working on long after others had retired to bed. Friends knew there was nothing strange about receiving a text message from him at 4.30 in the morning. Now the darkness was something he dreaded. He couldn't work, but couldn't sleep either, not for more than an hour or two. He felt exhausted, mentally and physically, but lay in bed with his mind racing, consumed by anxiety, listening to the traffic dying away on the Woodstock Road, the silence descending, and finally, after what felt like an eternity, the first liquid bird calls followed by a full-throttle dawn chorus as Oxford awoke. He went to see the college doctor, and asked for sleeping pills. He stopped going to the gym. A grey tide of depression had welled up and dragged him under, and there was no one on God's earth he could share his anguish with.

The crisis, a dark night of the soul that lasted two grim weeks, was a form of delayed reaction. However many times John might have noted in his diary that corruption in Kenya went to the top, his heart had never accepted what his brain told him. The *Mzee* could not be, must not be, the grand spider at the centre of State House's web of corruption. And now, having examined all the evidence, John knew that scenario made no sense. Kibaki was not out of the loop, deceived by manipulative aides, scattily ignorant of the system of sleaze operating all around him. He *was* the system. Kibaki and his cronies had played him for a fool, and he – star pupil, plucky former hack and experienced NGO wallah that he was – had kindly obliged. The bitterness choked him.

And with that came another terrifying realisation: 'This thing will never go away.' In his mind, up until then, John had managed to balance two parallel, if mutually exclusive, scenarios. Allow Anglo Leasing quietly to fade away, or clear his conscience and become one of the most famous – or infamous – Kenyans in history. What suddenly struck home was the understanding that it had to be one or the other, he could not have both. And it had to be the latter, with all it would involve in terms of public vilification and media hysteria, because of who he was. Character is destiny. 'John has the kind of honesty that stems from not being able to live with yourself if you don't do the right thing,' Ali Zaidi, his former editor, once told me. As a moral actor and a devout Christian, his route was virtually preordained. 'Initially, I never saw myself as a whistleblower. I had not thought it through to that point. Maybe part of me hoped all my work, my interactions with government people, would lead to internal changes that would be positive. But in the end I had to do the hard thing, the painful thing.' Travelling on the Oxford double-deckers, he gazed at other passengers and experienced a fierce pang of envy for their ordinary lives, their mortgages, their prosaic worries about which school to choose and whether they could afford a new car.

One dreary feature of this new life was clearly going to be an intimate acquaintance with various legal chambers. John already had a lawyer in Kenya, and as the months passed he would acquire additional lawyers in the United States and Britain. Thankfully, much of the work they did for him would be pro bono: he was the kind of high-profile client whose business added lustre to a chambers' reputation. Hanging over him was Kenya's Official Secrets Act. This catch-all legislation had long served as a curtain behind which government could conduct its affairs away from prying eyes. By chance, John had never been asked to sign it, although all his staff had done so. But that omission, in his own eyes, offered no real let-out – the keeping of state secrets had been implicit in his role. 'I can't in good conscience walk into a court and ask my lawyer to defend me on the basis that I didn't sign it.'

Before leaving Kenya, John had read up intensively on the Act, even asking a lawyer friend to draft a legal opinion. The question of whether it could be justifiable to divulge information acquired in the course of his professional duties – information clearly never intended for public ears – had haunted him since arriving in Britain. For John, this was not simply a legal issue, it was moral and spiritual. It bothered him so intensely that in the first months of exile he paid for an old friend, a devout fellow Catholic he had often prayed alongside, to fly from Nairobi to Oxford to serve as spiritual adviser. 'He was in a quandary,' remembers the friend. 'He wanted to know whether information of corrupt dealings that had only come his way because of his appointment could be used for purposes for which it was not intended. I went away, thought about it and told him that this information, which involved the stealing of public money, did not belong to the president or the government. It belonged to the *wananchi*. He was very scrupulous, he needed to be morally sure.'

Behind the Official Secrets Act lurked something even more alarming: a possible High Treason charge. High Treason carries the death penalty in Kenya, although such sentences are routinely commuted to life imprisonment. Crucially, it is a non-bailable offence, which meant that if John returned to Kenya he could be charged, immediately thrown into prison and – in a country notorious for the creeping pace of its judicial system and the squalid state of its jails – left there to rot virtually indefinitely.

One way to sidestep these problems would be to testify before parliament. The Public Accounts Committee (PAC) was a cross-party group led by opposition leader Uhuru Kenyatta, the man beaten by Kibaki in the 2002 elections. Anything said before the PAC would be privileged, protecting John from legal pursuit. He could not go to the mountain, so the mountain must come to him. If he could persuade Uhuru Kenyatta to bring the PAC to London, he would unburden himself before its members, satisfying his nagging desire to render account to the Kenyan people. When, in August, planning minister

Peter Anyang' Nyong'o raised the possibility of John testifying before the PAC, John responded with alacrity. He was ready to talk, he declared in his first press statement. He was simply awaiting a government invitation.

The government reaction was a telling silence. It was the same story when John, hearing that Kibaki would be in London to meet East African investors, once again allowed himself, for a moment, to pin his hopes on his former boss. 'I've been in touch with State House, suggesting that I present my evidence directly to the old man at the High Commission. I owe it to him,' he told me. 'If you do that, make sure someone else – me, Michael, anyone – goes with you,' I emailed, suddenly convinced that if John entered the Kenyan High Commission in Portland Place alone, he would never come out. I needn't have worried. The president had no desire to talk to his troublesome former *kijana*, and John's offer was not taken up.

One of the elements miring him in his Slough of Despond was the grim isolation of his position. 'When you effectively blow the whistle at that level, there's no one you can speak to. There's no one who can tell you what it's going to be like, the opprobrium that is going to result, no one. There are no precedents.'

Back in Kenya, Joe and Mary Githongo had withdrawn into their Karen villa, aware that many old acquaintances hesitated to be associated with a family which had produced such a wayward son. Their phone line had become so crackly, conversations were virtually impossible. Mrs Githongo had gone to complain at the Telkom Kenya office, but thought she knew the reason for the interference: the line was being monitored. There had been an unnerving episode when a Kenyan MP turned up at the residence – which Joe Githongo had used as collateral for a bank loan – claiming he had heard it was for sale. Two flustered, vulnerable old people were being used to send John a message.

When I visited John's parents in Kenya that August, I had a sense of a couple disconcertingly ignorant of what was to come, but hunkering down in cautious preparation for any eventuality. With his sons' help, Joe was trying to settle his debts and sort out his tangled business affairs. Mrs Githongo, for her part, had sold a stretch of the land at the back of the plot where the family had until recently run a small farm. She planned to do the same with several plots upcountry. Walking slowly, to allow an adored granddaughter to trot alongside, we inspected the empty chicken runs and livestock pens, which were being clucked over disconsolately by John's aged wet-nurse and one of his aunties. 'You know, Kikuyus and land, you never let it go,' said Mrs Githongo, observing the women's distress. 'But look at Kenyatta, look at Moi – so much land, for what? Let others enjoy what we have.'

The couple tried to keep their spirits up, tracing Christian parallels in their son's tribulations. 'Jesus came from Heaven and had to die for us all,' said Mary Githongo. 'Someone had to sacrifice his life for others.' But they had no real grasp of how great that sacrifice was likely to be. John, they told themselves, would return once the crisis had blown over. Let him do a Masters, maybe take a job with the African Union, or go to work in the States. It was just a question of waiting things out.

John's brothers, with whom he exchanged constant emails, were far more aware of what was in the offing, and nervous with it. His departure had altered the shape of their lives, too. Younger brother Mugo had also noticed problems with his phone. He tried to keep a step ahead of his shadows by regularly buying new SIM cards, switching email accounts and changing the locks to his apartment. But a low profile felt like the best precaution. This was no time, he said, to go drinking at night in Nairobi's bars. 'My social life has stopped. I'm very wary. I don't go out.' It was too easy, in a city notorious for violent crime, for a political hit by someone wanting to get at John via his family to be camouflaged as a random mugging. Mugo never wore his seatbelt in the car these days, he let slip, because he wanted full freedom of movement in case of sudden incidents on the road ahead.

Gitau also seemed in sombre mood. We met for a beer next to the swimming pool of my one-star hotel. Evening was drawing in and the first bats were darting out from the palm trees as the light died, skimming the pool for midges. Although he was in fact the second Githongo son, Gitau was often mistaken for the oldest sibling. He had the same bull-neck, solid jaw and sheer heft as John. He was the one who had got lumbered with the family firm, and if Mugo was the T-shirt-wearing radical puppy of the family, Gitau, grave and balding, looked the accountant he had become, conservative and cautious. He had many friends in the business community, and they took a distinctly pragmatic view of Anglo Leasing, he told me. As long as they were making money, they could tolerate sleaze. 'They're telling me: "In whose interest is it for the government to fall? Let these tenpercenters have their 10 per cent, what we care about is stability." What you have to realise is that Kenyans don't really believe in democracy.'

If he fretted about the explosive impact of his brother's revelations, Gitau also worried, quixotically, about the precise opposite. 'What if John spills all, everything he knows – and *nothing happens*?' The media in Kenya were increasingly complicit, the political class supine, and what did the donors' reaction – if there was any – matter when China stood ready to lend to African governments, no questions asked?

We walked together to his car, a classic old Mercedes the colour of dull gold. Before saying goodbye, Gitau halted and turned to pose a question that had clearly been preying on his mind.

'Has anyone else ever done this?'

'Watergate?' I suggested.

He shook his head. 'I mean in Africa.'

I thought for a bit, but couldn't recall a single occasion in which a government official of John's stature had blown the whistle on an African administration. There were no examples pointing the way ahead. Gitau nodded gravely, a worry confirmed, and got in his car.

By early September 2005, John had compiled a ninety-one-page, 40,000-word dossier. Should Anglo Leasing ever come to trial, it was the kind of document any prosecutor would fall upon with cries of appreciation, half his job done for him. Reading, in many ways, like a real-life thriller, the dossier followed the course of John's growing enlightenment chronologically, incident by incident, conversation by conversation, weaving in quotes from his diaries, citing numbered tape recordings and supporting official documents. He would keep adding nuance and detail to this report as the months went by, but the bulk of his evidence was now in place. In his own mind, John had finally crossed his 't's and dotted his 'i's.

Yet while he'd been working on this dossier, events back in Kenya conspired to flummox him.

In mid-September, Kenya's electoral commission named 21 November 2005 as the date for the country's long-promised referendum on a new constitution. What Kibaki had pledged to deliver in the first hundred days of his presidency had taken nearly three years, and the final version was a world away from what the Kenyan people had once envisaged. It was true that on issues such as land ownership, women's inheritance and the role of religious courts, the proposed constitution offered radical change. But in the eyes of its critics – who happened to include six members of the NARC cabinet – it failed to deliver on the critical issue that had blighted politics in Kenya since independence. Whereas the first Bomas draft had proposed dividing executive powers between a president and an executive prime minister, the final version included a non-executive prime minister, subservient to a still-supreme president. For Raila Odinga, this was the ultimate betrayal. Once again, the wily Kikuyu had shown that they could not bear to share. The Memorandum of Understanding signed on the eve of the 2002 election had been violated, the faultline at the heart of a hastily assembled coalition exposed. And if much of the rest of the country felt distinctly nervous about the idea of Raila as prime minister, they were far from happy at NARC's sleight of hand. A historic opportunity to place the country's system of government on a more equitable footing had been missed. Both the cabinet and the country divided along ethnic lines, with the Kikuyu, Meru and Embu rallying behind a 'Yes' vote, symbolised by a banana, while a majority in every other community called for a 'No', represented by an orange.

These were violent times in Kenya, as the Orange and Banana camps clashed on the campaign trail. Bombarded with information from his sources, John felt pulled this way and that. This, he finally decided, would be the worst possible time to come out with a corruption dossier. If he went public before the referendum, it would be seen as a blatant political move, aimed at boosting the Orange campaign. It was frustrating, but he did not want his dossier reduced to campaign fodder. He stowed it in a safe deposit box and prepared to wait the referendum out.

Unaware of this decision, the Mount Kenya Mafia extended an agitated feeler. Lands and settlement minister Amos Kimunya and Dr Dan Gikonyo, Kibaki's personal physician, turned up in Oxford to negotiate a quiet understanding. A smooth-talking Kikuyu, Kimunya was regarded by diplomats as a representative of a promising breed of young statesmen rising through the ranks in Kenya. A US-trained Kikuyu cardiologist, Gikonyo was a doctor with a political profile. He had always been close to the Democratic Party, patching up opposition activists beaten by security forces during the Moi years, and had been constantly at Kibaki's side since his near-fatal campaign car crash. His practice had thrived, and he was about to open a 102-bed, four-storey private hospital in Karen, boasting state-of-the-art scanners and TVs in every room. Coincidentally, Gikonyo was also physician to Joe Githongo and, via that association, to John himself. That no doubt explained why he had been sent with Kimunya to woo John – who but a priest can rival a doctor for leverage over a trusting patient?

They booked a table at Brown's, one of Oxford's most popular restaurants, a few minutes' walk from St Antony's. The evening started cordially, with broad smiles all round, but deteriorated when the emissaries began delivering their message. As voices rose, the waiters exchanged glances and lifted eyebrows, wondering whether the evening might end in blows. Kimunya and Gikonyo were there to make sure John did nothing to blow the referendum campaign off course. 'They kept saying, "SWEAR to us, SWEAR that you won't spill the beans before the referendum. You must swear, John."' Sensing resistance, Kimunya made the mistake of appealing to John's supposed ethnic loyalties. 'Do you really think uncircumcised people can rule Kenya? We are going to sack all these Jaluos in government, replace them with other Jaluos and Kenyans will just forget.' Kimunya had just encapsulated the thought process – that Kikuyu assumption of a divine right to rule – that repelled the rest of Kenya. 'These were guys who went to university,' says John, 'educated people with international experience, not uneducated villagers from the sticks, talking like that. I lost my cool, I admit, I got very emotional.' Kimunya followed up the crude tribal rallying cry with a stark reminder of the reality of Kenyan politics. Break your silence, said the minister, and 'Your grandchildren will regret.' It was a traditional Kikuyu way of saying, 'If you don't cooperate, there won't *be* any grandchildren to succeed you.'

The encounter left a sour taste in John's mouth. With the referendum less than a month away, it was too late now to spring into action. But he hated the sense that, through his inaction, he had played into these men's hands. 'I felt very angry. I said to myself, "What have I done? I've quit the stage and left it to these buggers."'

The delegation to Oxford might have got what it wanted, but it made no difference to the referendum result. Nor did the vast sums of stolen Anglo Leasing money spent attempting to secure the vote. Referendum day became a poll on the very principle of Kikuyu rule. John stayed up to monitor the various Kenyan newspaper websites updating their results throughout the night. To his delighted amazement, Kenyans showed that while they were willing to be paid, they could not be bought. In the privacy of the polling booth, they cheerfully voted against those who had bribed them. Over 58 per cent rejected the new constitution consolidating the presidency's supremacy. Out of eight provinces, only one – Central Province, GEMA's heartland – voted 'Yes'. The country had delivered a stinging slap to an ethnic group whose leaders believed themselves born to rule. The text messages from excited friends back in Kenya came so thick and fast, John's mobile gave up the ghost. 'I think it just melted. It couldn't take any more,' he chuckled.

A devastating rebuke of ethnic conceit, it was an appalling result for the Mount Kenya Mafia, and one that caught them unprepared. Kibaki immediately dissolved his fractured cabinet and suspended parliament. John gave one of his barrel laughs when he began receiving text messages from desperate government officials begging him – *him* of all people – for advice. A siren call came from one adviser in Nairobi, a seasoned political observer. If ever there was a chance for a leader to learn the error of his ways, he argued, this moment between the dismissal of the old government and the appointment of a new one was it. Punished by the electorate, surely Kibaki would recognise he had fallen into bad company, ditch the Mount Kenya Mafia and open his arms to true reformers? John fell for the old seduction one last time. 'I was excited, although my gut told me not to be. He'd sacked his entire cabinet, this was an opportunity.'

The day after the referendum, John distilled his ninety-one-page dossier down to a thirty-six-page summary. Tailor-made for a man with a packed diary and a short attention span, it was a document that could be digested in less than an hour and a half. 'This is my statement of events leading up to my resignation in February 2005 . . .' it began. Addressed to 'Your Excellency', it was full of poignant reminders that none of what was being described was in fact unfamiliar to the person for whom it was intended, reminders that in themselves should have alerted John to the futility of the exercise. 'I briefed you on all these matters . . .' he wrote, 'I briefed you on this . . .', 'I updated you . . .' The mini dossier was effectively the last of his many presidential briefings. 'Here, in black and white,' John was telling his boss, 'is what I know, what I told you and what you should have acted upon – and still can.' Crucially, it contained no reference to the existence of any tape recordings. But then, Kibaki had already been told about those long ago, so why repeat himself?

Having dispatched the dossier, John used his informers network to reach into the entrails of State House, monitoring each stage of its progress. Even now the old Kibaki magic held him in thrall – before he could go any further, he needed to *know*, with absolute certainty, that the president had read the dossier. The whispers kept him abreast. The dossier had been placed on the president's desk. The president had spent an entire afternoon reading it. The dossier had been replaced on the desk without comment. Silence.

On 7 December 2005, Kibaki named his new government. With the sole exception of Chris Murungaru, all the key ministers associated with Anglo Leasing were reappointed to cabinet. There had been no attempt to heal the wounds opened by the referendum or to distance himself from his administration's biggest scandal. Instead, those who had led the Orange campaign, including Raila Odinga, were summarily ejected. Stuffed with trusted Kibaki cronies, the government could no longer be described as a broad-based coalition. It had become a narrow expression of Mount Kenya power, with a few opportunistic ethnic outsiders tagging along for the ride. Far from seizing an opportunity to reform, General Coward had withdrawn even further into his ethnic citadel.

Failing to win a response from Kibaki, John had also sent a copy of the dossier to Justice Ringera, asking to be invited to Kenya. 'I'd told him, "I'm ready to come and share incontrovertible proof with you, just ask me."' It can't have surprised him when the head of the KACC, the man who had helpfully passed on threats to his life, failed to respond. '"That's it," I thought. "There's no one else to talk to now."'

One last option remained – the media. His depression lifted. A born hack, he always felt at his best when things were on the move, when he had a defined target and a pressing deadline. After the long months of waiting, he felt a surge of adrenalin. I could gauge his gathering excitement in a growing tendency to resort to journalistic cliché. 'I'm keen to get this monkey off my back,' he told me.

14

Spilling the Beans

'It always seems impossible until it is done.'

NELSON MANDELA

Headquartered in one of Nairobi's most eccentric buildings – a zebra-striped, twin-pillar folly which bears more than a passing resemblance to a giant liquorice allsort, the *Daily Nation* is not Kenya's oldest newspaper. It is, however, its largest and its best, flagship of a vibrant media group whose radio studios, television stations and newspapers are sprinkled across Uganda, Tanzania and Kenya. Curiously, East Africa's version of Rupert Murdoch is the Aga Khan, spiritual head of the Ismaili religion, whose influence over the region is quirkily out of kilter with the modest size of the community he represents.

Under Moi, the *Nation* was a must-read for any thinking Kenyan. When cautious diplomats buttoned their lips and nervous international lending institutions kept their mutterings private, it often seemed that only the *Nation* had the guts to denounce the latest abuses. But when NARC came to power, the *Nation* lost its moorings. Top management, heavily Kikuyu, was close to the former opposition leaders now running the country. 'The message from the NARC government was: "Look, we're on the same side. You want change, we have an agenda for change, give us a chance,"' a *Nation* editor told me. New instructions came from the top: in future, there were to be no more scoops quoting anonymous 'government sources'. If the sources couldn't be named, the story would not be printed. The instruction effectively closed down most reporting on government sleaze. One Western diplomat in the habit of leaking titbits noticed the change. 'Even the good guys were scared stiff. They wouldn't publish anything unless I could provide documents to back it up. And things that would have been really easy to follow up just weren't.' Readers noticed that their favourite sharp-tongued columnists were either leaving the newspaper or wrote far less. 'These days you have to read the *Standard* to know what is going on,' a friend told me. 'You just can't trust the *Nation*.' So how would management react when John, one of the Nation Group's former columnists, dangled the scoop of the decade under its nose?

The link with the *Nation* was one of the many relationships John had been careful to maintain since leaving Nairobi. It had flared into vibrant life after the referendum, whose outcome had demonstrated how out of touch the *Nation*'s board was getting with public sentiment. Falling circulation figures showed Kenyan readers were snubbing a newspaper increasingly perceived as a Kikuyu-dominated government mouthpiece. In mid-December, Joe Odindo, the Nation Group's managing editor, confirmed that he and editorial director Wangethi Mwangi had persuaded their bosses to send them to Oxford to find out what the anti-corruption chief had to say. 'The train has left the station,' John told me.

For John, those febrile four and a half days together were as cathartic as any Catholic confession. 'All this time, you know, except for you and Michael and a couple of others, no one has actually known what happened,' he told me. Unburdening himself was not only a massive relief, it allowed him to confirm the significance of his material. Compiling the dossier, John had become numbed to its intrinsic shock value. Now he watched the faces of the two journalists, hunched in their headphones in his study, as they followed his emotional itinerary, their expressions moving from perplexity, through disbelief, to horrified acceptance. 'These were veteran journalists, very hardnosed guys, and they were stunned to hear guys openly discussing stealing money.'

The fight against corruption had always been handicapped by a lack of evidence, he told them. This was no longer true, and the *Nation* was being given the chance to scotch a pattern of behaviour that dated back to independence. 'If we can strike a blow against that way of doing things, confuse it, disorientate it, I see that as a major contribution.' Breaking the story, he argued, was a patriotic duty. The continent's regeneration could not be left to outsiders, whose low expectations of Africa were part of the problem. 'It's our country. Let's do this. If we act on the premise that, "Oh, this is African politics," then it's a self-fulfilling prophecy.'

Putting in fifteen-hour days, interrupted only by short walks along North Parade to clear their heads, the three planned a fortnight's worth of coverage. The Kenyan public, the newsmen agreed, would be hit by day after day of devastating exposures. 'It can't just be one set of headlines. We must rub it in, and rub it in endlessly,' urged John. He had essentially given them the story, but the *Nation* still needed to build it, block by block. In view of the Official Secrets Act, the dossier could serve as no more than a backbone, a skeleton to which the *Nation* would attach flesh, sinew and skin. The paper's chafing battalion of journalists must be let loose to dig out the quotes, anecdotes and details that would not just corroborate John's tale, but take it to a higher level.

Carrying the dossier with all the reverential care of two Knights of the Round Table bearing the Holy Grail, Wangethi and Odindo flew back to Nairobi and immediately on to Mombasa, where a member of the board was spending his Christmas holidays. 'We must do this, or we are finished,' was his reaction. The managing board and the Aga Khan swiftly concurred, and a team of reporters was assigned. Time was of the essence. State House was certainly aware of Odindo and Wangethi's trip to Oxford, and would soon, thanks to its contacts on the newspaper, learn of the *Nation*'s plans.

Braced for the storm, John suddenly received a blow from the most unexpected quarter. For months he had been discussing what was to come with Mary, his girlfriend. Yet clearly she had never really taken in his words. Now, belatedly, she panicked and alerted his parents, who tried to stop the speeding juggernaut. 'You'll destroy the family firm,' said Joe. It was far too late, and Mary's loss of nerve was a setback from which the relationship would never totally recover. 'It's like someone using a walking stick, and the walking stick is suddenly knocked away,' recalled John. 'I had no words. It was huge.'

The story finally broke in the last week of January 2006, days before the first anniversary of John's departure. Having tried so carefully to stage-manage the scoop, John found his plans sabotaged by messy reality. Aware that the story was about to break – he had been sitting on a copy of John's dossier for weeks, after all – KACC director Justice Ringera did his best to neutralise its impact by cannily announcing an impressive-sounding blitz of Anglo Leasing-related interrogations. Those called for questioning included vice president Moody Awori, finance minister David Mwiraria, justice minister Kiraitu Murungi and former internal security minister Chris Murungaru. Simultaneously, an officer at one of Nairobi's biggest embassies invited a handful of British journalists to read a copy of the dossier that had been quietly circulating in diplomatic circles. Obliged to double check every allegation, the *Nation* had taken too long, allowing itself to be scooped by a recently-arrived Western reporter who had never even met the anti-corruption chief, Xan Rice of *The Times*.[34]

John would in future be constantly berated for making his revelations in the international rather than Kenyan press, a bitterly ironic accusation given the lengths he had gone to to provide the *Nation* with an exclusive. But the deed was done, and its impact was not to be denied. 'It was,' in the words of US ambassador William Bellamy, 'like a grand piano falling out of the sky.' The *Nation* released its meticulously prepared stories, and its rivals and every Kenyan radio station buzzed with the details John had so long kept to himself: the regular briefings with Kibaki, the death threats, the attempts by colleagues to block his work, their open admissions of guilt.[35] Opposition MPs called for the government to be dissolved, civil society and the Catholic Church demanded top-level sackings. In one of many surreal attempts at face-saving, lands minister Amos Kimunya, the very man who had gone to Oxford to beg John's silence, issued a statement inviting him back to share his evidence. This was somewhat undermined by the leaking of a report commissioned by the government, recommending that John be charged with high treason upon arrival.

The first head rolled on 1 February, with the resignation of David Mwiraria, the finance minister. Was it the first time in Kenyan history a minister had resigned over corruption? Most people thought so. By then I was in Nairobi. 'Congratulations on your first scalp,' I texted John in Oxford. 'The first one bites the dust,' was his reply, hinting at a previously unseen streak of vindictiveness. Others proved harder to dislodge. Revealing the hard core of self-interest beneath his cuddly image, vice president Awori announced he was staying put. As Kiraitu, too, dug in his heels, John pulled a second pistol from its holster and took aim. On 9 February BBC World broadcast a long interview between John and Fergal Keane in which his recording of the Kenyan justice minister's clumsy blackmail attempt was played. He had initially offered the tape to the Nation Group, but they had not had the courage to broadcast it. Since the Kenya Broadcasting Corporation routinely relays BBC World's programmes, the interview went out on prime-time television, not once, but repeatedly. Later that evening, viewers were treated to the spectacle of Kiraitu Murungi scampering down a corridor as a female correspondent chased after him, shouting, 'Mr Minister, Mr Minister, are you going to resign?' On 13 February he too agreed to 'step aside'. With him went presidential aide Alfred Getonga and George Saitoti, the education minister whose name had been repeatedly mentioned in connection with the Goldenberg scandal.

Withdrawing into near-total silence, Kibaki was jettisoning ministers like ballast in an attempt to keep his balloon aloft. The message of the joint resignations was clear: if I'm going to be punished for Anglo Leasing, then my predecessors will be punished for Goldenberg. It had taken thirteen years for a minister to resign over Goldenberg, but only a few weeks for heads to roll over Anglo Leasing. Perhaps some things in Kenya were speeding up.

For those in government, the discovery that John had taped his conversations felt like a heart-stopping kick to the stomach. Remembering overly frank exchanges, they felt the nausea of real fear. *Just how much did he have?* For ordinary Kenyans, the revelation came as a hilarious moment of liberation. Accustomed to the brain-numbing exchange of accusation and counter-accusation that followed every corruption scandal, they could hardly believe someone had thought to collect the proof. 'Is this true? Is this true?' asked my driver. 'Did he tape them?' 'Yes, that's what he did.' He slapped his hand on the steering wheel, marvelling at John's ingenuity. 'Oh, this is a man! This is a straight man!'

Commentators who had dismissed John as a sell-out revised their opinion. 'I used to criticise John Githongo for being a vain man, a naïve man, in accepting that job,' acknowledged columnist Wycliffe Muga. 'I was wrong. What he has done shows that he was, in fact, the right man for the job.' Over their lunches, office workers speculated whether Githongo had gone the whole hog and done the unspeakable: taping the president himself. Pundits joked that in future civil servants and politicians would have to attend meetings naked, in order to ensure no one was wearing a wire. Everyone had an opinion now, for the BBC had posted the thirty-six-page Githongo dossier on its website. A report originally written for presidential eyes could now be downloaded and pored over at any internet café. John had acquired the status of a Grand Confessor, implacable arbiter of truth. In Nairobi bars, the occasional drinker, growing belligerent over his fourth beer, could be observed wagging a finger at a friend and pronouncing: 'I tell you, I would say this *even if Githongo were here.*' In western Kenya's opposition stronghold, there were reports of newborn babies being baptised 'John Githongo'.

Retribution was something Kenyans had not experienced before, and they found they liked the flavour. Having tasted three ministerial resignations in a fortnight, they hungered for more. In the foyers of offices and hotels, staff gathered around television screens, watching their discredited ministers attempt to justify themselves before sceptical interviewers. 'It's getting better,' my newsagent told me, pointing to the resignation headlines on the newspapers he was handing over. 'But we are expecting more. Now it's time for the vice president and attorney general to go.'

The reaction was not overwhelmingly approving, however. 'The image of a traitor who fled to enemy territory to abuse his motherland can't quite leave my mind. I could feel something clogging my throat,' wrote one *Nation* reader. 'No matter how useless we are as a country, can't we be spared this international embarrassment?' Columnist Edith Macharia warned John in the same newspaper not to 'expect any laurels for airing the country's dirty laundry'. 'I fear he will go the way of all informers and snitches – ignominiously,' she sniffed. It was noticeable that the angriest attacks on John came from Kenyans with Kikuyu names, like Macharia, appalled that a kinsman should have exposed the House of Mumbi to such ridicule. The real vitriol was reserved for the Kikuyu websites, where members of the community tore into one another. While some hailed John as 'God's gift from heaven', others accused him of running into the arms of the *mzungu*. 'Enjoy your BBC moment, 'cos you're done for, Githongo,' wrote one. 'May you rot in hell.' Wycliffe Muga was not alone in offering me some quiet advice. 'If you are one of John's friends, tell him not to come back. If the Kikuyus found him they would kill him without even waiting for instructions from above. They are very, very bitter.'

Having claimed his first scalps, John barely paused for breath. Even before Kiraitu had accepted the inevitable, John had travelled down to London to testify before the Public Accounts Committee, whose members had flown to Britain to hear his evidence. Entering the Kenyan High Commission on Portland Place, John effectively had his longed-for moment in court, testifying for two solid days.

So far, much had gone his way, but emerging from the High Commission on the final day of his evidence, John was reminded of the perilousness of his course. He had always known the revelation he had taped his colleagues would revolt as many Kenyans as it impressed. Some of the Kenyan journalists who gathered there that day were drunk, and they hurled the familiar accusations in his face.

'Did you tape the fucking president?' shouted one. 'Did you have a gay relationship with Edward Clay?' called another. 'Who's paying your fees?' Dickson Migiro, a young Kenyan television producer there that day, cringed at the spectacle. 'I thought, "My God, is this Kenyan journalism?", because they were chasing the guy down the street.'

In mid-March John was back at the High Commission, this time to testify before Ringera's KACC. Just as a government delighted to see the back of him had been forced to invite John home, circumstances now required the judge who had told him to stop pursuing Anglo Leasing to go through the motions of quizzing him.

That exchange marked the slow pulling out of the tide. It would take John a while to realise it, but his testimony before the Public Accounts Committee marked the high-water-mark of his campaign for accountability. Plotting his course, he had divided his operation into seven separate stages. Stage One was the exposure of the system of looting. Stage Two weathering the public reaction. Stage Three his testimony before the PAC and KACC. Stage Four would be the bringing about of political change, however small. The next stages involved the return of stolen monies and prosecution – the last of which, while a theme harped on by opposition and human rights groups, seemed to John a virtually hopeless cause in a litigation-prone nation. 'These guys own the judiciary, they've stolen the money and can pay the lawyers to fight for them for years. The crucial thing isn't prosecution, it's restitution, payback.'

In the event, Stages Five, Six and Seven would slip forever from his grasp, retreating into the nebulous realms of What Might Have Been. The coming year would see a gradual, systematic clawing back of lost ground by the scandal's perpetrators and their defenders. So formidable would that campaign prove that even holding on to the rewards of Stages One to Three would come to seem something of an achievement.

15

Backlash

> 'A seventy-year-old grandfather stunned a Nyeri court yesterday when he admitted that he wanted to commit suicide, following disappointments caused by the political turmoil in the country. Stephen Nyamu Ngari said he wanted to escape the political wrangles witnessed since the introduction of multi-party politics. He caused laughter when he said that he was tired of life under the current political order. The magistrate ordered that he be taken for a mental check-up and be produced in court on Friday.'
>
> *East African Standard*, 19 April 2006

Soon after midnight on 2 March 2006, staff working the night shift in the Kenya Television Network headquarters in Nairobi's I&M Bank Tower, a gleaming skyscraper that pierces the skyline like a splinter of blue ice, got a nasty shock. Viewers of KTN, part of the Standard Media Group, would later be able to see exactly what happened, thanks to CCTV footage replayed on national television. Commandos in balaclavas and hoods, wielding AK47s, broke into the transmission room where two technicians were watching a row of screens. Driving up to the I&M Tower in unmarked cars, the raiders had roughed up nightwatchmen and smashed their way through a series of locked doors to get that far. Ordering KTN staff to close down the station, they kicked them to the ground and then proceeded to ransack the offices, removing vital broadcasting equipment, scores of hard drives and computer monitors. Only solid doors saved the *Standard*'s editorial floor from similar damage. Frustrated, the hooded men roared off to the *Standard*'s printing plant in Nairobi's industrial area, where they smashed machinery and made a bonfire out of thousands of copies of the next day's edition.

Smarting from weeks of merciless media attack, the Mount Kenya Mafia had struck back. If the *Standard*, Kenya's second-biggest newspaper, had not led domestic coverage of the Anglo Leasing affair, it had joined in with relish. While a certain amount could be forgiven the *Nation*, in light of past friendships, Moi's ownership of the *Standard* meant any criticism from that quarter was viewed as politically charged. 'Your newspaper is dangling by a thread,' Alfred Mutua, the government spokesman, warned Pamela Makotsi, the *Standard*'s editor, the day before the raid. 'The government is looking for the smallest excuse to slam down on the *Standard*. They can raid you, you know, just like they did in the past.' She had been incredulous. 'What, this new government, raid us? Come on, that's what happened in the old days.'

It took a while for the government to admit responsibility for an operation whose brute thuggery seemed more typical of Mugabe's Zimbabwe than of Kenya. But when it finally did, there was no apology. The *Standard* had been planning a story that posed a threat to national security, the government claimed. 'If you rattle a snake, you must be prepared to be bitten,' said John Michuki, minister for internal security.

The raid, which came as John was preparing to give testimony to the KACC, had several curious characteristics. One was that it had been staged without the prior knowledge of police commissioner Major General Hussein Ali. The other was that the technicians ordered to the floor had noticed, despite the balaclavas and hoods, that several of the commandos were white. It was the first, but not the last, controversial appearance of the 'Artur brothers' – two gun-toting, rule-flouting, medallion-sporting, party-throwing, bling braggarts whose sojourn in Nairobi would deliver a sharp kick in the groin to what little remained of NARC credibility.

They called themselves Artur Margaryan and Artur Sargasyan, boasted that they were descended from Armenia's former royal family and related to the Armenian prime minister. It later emerged that their passports were stolen, their real names a mystery, the Armenian premier had never heard of them, and their nationality was just as likely to be Russian or Czech. With their black shirts, heavy gold chains and close-cropped heads – which only served to accentuate the contrast between their pallid East European skins and dark hair – they certainly looked what opposition leader Raila Odinga claimed them to be: the least discreet of international mercenaries.

Raila claimed government operatives had hired the two not only to lead the raid on the *Standard*, but to assassinate opposition politicians. Not true, replied the two Arturs, they were harmless international investors who would bring prosperity to Kenya via – now, where had Kenyans heard this before? – their role in Africa's gold and diamond trade. But everything about the two Arturs was off-kilter. Strangely talkative, they staged the first of their press conferences in Jomo Kenyatta airport's VIP lounge, an area usually inaccessible to ordinary mortals. Reports circulated that they were part of a Russian mafia network and had been sent to Kenya to 'liberate' a massive cocaine shipment seized at the coast in 2004, the biggest haul in African history. When they were at last deported in June 2006, having finally crossed the line by waving a gun at customs officials who tried to open their luggage at the airport, their villa was found to contain a collection of assault rifles, bulletproof jackets, a small fleet of cars, and official documents assigning them the rank of deputy police commissioners and giving them unrestricted access to all Kenya's airports. Decidedly, these were unusually well-connected soldiers of fortune.

Through his various security contacts, John heard that the Artur brothers had been spotted in Britain. The freelance security experts' assignment, he was told, was twofold. Their first task was to install listening devices at the Kenyan High Commission in London in order to record his closed-door testimony before the PAC. The second was to rid the Kibaki presidency, in the style of Henry II, of this turbulent whistleblower. The Thames Valley Police had heard similar reports, for after a six-month silence they suddenly got in touch. 'It was very discreet, very gentle. They just said: "We've heard about the Russians. You might want to move."' John swapped his exposed lodgings on the Woodstock Road for a room in the womb of the college campus, and was issued with a set of alarms.

The devices were never put to the test. Did John's enemies lose their nerve at the prospect of carrying out an assassination on British soil, an act that would have plunged Kenya into the same kind of diplomatic and political hot water as the Soviet Union in the wake of Alexander Litvinenko's fatal poisoning later that year? Did the Artur brothers, so clod-hoppingly indiscreet in all their operations, simply fail to deliver? John could only guess. The brothers were tracked as far as Oxford. 'It was incredibly incompetent. These guys were spotted drinking downtown within days.' And then they disappeared from the radar, returning to Kenya mission unaccomplished.

The episode had provided John with a timely reminder of how far his enemies, facing a forthcoming election, were ready to go to stay in power. But the entire Artur saga carried wider lessons for anyone concerned about Kenya's trajectory. One was that the president's entourage was careering out of control. The other was even more worrying. The brothers' swaggering contempt for the regulations suggested they thought they had landed in a banana republic, where those with friends in high places need fear neither the journalist's microphone nor the policeman's knock. It was a view they increasingly shared with Kenya's foreign donors, in whose eyes the government was beginning to look about as respectable as those of Colombia or Albania.[36]

For the United States in particular, whose embassies in Nairobi and Dar es Salaam had been bombed by Al Qaeda in 1998, the Arturs affair highlighted the vulnerability of a society eaten away from within. Corruption can reach a point – and it seemed, with the Artur saga, that point had been reached – where an entire nation is there for the taking, where sleaze has security implications not just for the nation concerned, but for its neighbours and allies. 'The reason Al

Qaeda came here in the 1990s wasn't for the scenery,' Ambassador Bellamy told me. 'If you can lie your way through immigration, if you can get your goods through customs, if you can induce law enforcement to turn the other way, if you can sort out your legal problems with a few attorneys and judges, if you can launder your money and invest in legitimate businesses, well, why *wouldn't* you come to Kenya?' And if Al Qaeda found Kenya a congenial environment, thanks to the ease with which officialdom could be bought, then so did global criminal syndicates, drug traffickers and warlords wanted by international war crimes tribunals.

In the final stretch of his own posting, Bellamy had reached the same stage of hair-tearing exasperation as Edward Clay before him. 'The raid on the *Standard* was the nadir. It's like watching a Greek tragedy, in which the protagonists are doomed from the start to take the wrong stand at every turn.' The country had become a land of opportunity for the international underworld. Later that year, British Foreign Office minister Kim Howells revealed that Kenya had become a major transit point for drugs traded on British streets, with nearly a dozen domestic drugs seizures proving to have Kenyan links.

Those at the top had bent the rules so often they could no longer tell the difference between the legal and the illicit. Already, worrying new scams were pushing the Anglo Leasing scandal into the shadows. The latest involved Charterhouse Bank Ltd – funny how Kenya's conmen, true to the country's imperial history, always favoured names redolent of public school and playing field – a Nairobi bank suspected of laundering $1.5 billion-worth of illicit proceeds. Like the Artur brothers, the institution seemed to enjoy top-level protection. When the Central Bank moved to investigate it, its governor was suspended by finance minister Kimunya, its investigators accused of nepotism, and two whistleblowers who had alerted the authorities promptly took up an offer of asylum in the United States. Shopping in Nairobi that spring, I ran into a senior diplomat who told me his posting would shortly be coming to an end. Was he sorry to go? I asked. He shrugged. Well, yes and no. His embassy was very concerned about this Charterhouse business, a scam which seemed to dwarf even Anglo Leasing. 'There's a sense that every time you lift another stone, you find more things squirming underneath,' he said. 'It just never seems to end.'

In the light of the devastating revelations of top-level complicity, given the mounting evidence of State House's links with global gangsterism, what action would the country's donors decide to take following the leak of John's dossier? The answer was complicated. Kenya was about to find itself at the heart of a global ideological debate over corruption and development, a battleground between aid idealists and aid sceptics, the practitioners of *realpolitik* and those with dreams of saving the world.

Makhtar Diop had been replaced as World Bank country director for Kenya by one Colin Bruce, a Guyanese. Bruce must have known that his predecessor's tenancy had sparked controversy in Kenya, but he moved into the house inside president Kibaki's compound in Muthaiga notwithstanding. 'I told him not to do it,' one of his Kenyan acquaintances told me, with a shake of the head. 'But he wouldn't listen. These World Bank guys revel in that intimacy. They want to be part of that class, to hang around those people.' When asked about his living arrangements by journalists, Bruce refused to comment. He set about forging even closer ties with his landlord and key cabinet members than those enjoyed by Diop. 'He rationalised it by saying the government was not a uniform entity, there were the old guys and the reformists, and it was crucial to align with the reformists,' recalls a former colleague. Those reformists, in Bruce's mind, included finance minister David Mwiraria, who, he assured colleagues in Washington when questions arose about the politician, was 'totally clean'.

Demonstrating a truly remarkable sense of timing, the World Bank chose to announce $145 million in new loans to Kenya – the first credits approved by the executive board for fifteen months – just three days after the leaking of John's dossier, signalling that, as far as this institution was concerned, a $750-million procurement scandal was no grounds for querying the wisdom of re-engaging with the

Kenyan government. The same emollient message came from DfID, which had announced a £58-million grant a few days before John's leak, and saw no reason to reconsider. The IMF was more wary, refusing to release two scheduled tranches of a lending programme, and a consultative group meeting between the Kenyan government and its donors, originally planned for April, was put on hold. But the Netherlands was the only bilateral donor to announce it was actually freezing aid over corruption concerns.

How did these organisations manage to persuade themselves the Githongo dossier was of such limited relevance to their funding programmes? By taking the long view and fixating selectively on statistics. 'The trajectory is positive, the direction of travel is correct,' aid officials chorused when I interviewed them in Nairobi, pointing to GDP growth averaging 6 per cent over four years, such a relief after the minus scores of the Moi administration. Of course Anglo Leasing had been very depressing, but – and here they began to sound worryingly like NARC politicians – it was important to remember that more than half the eighteen suspect contracts had been signed under the previous administration. Perfectionism was pointless. Kenya was a graft-blighted society, certainly, but it was improving, and by encouraging the establishment of institutions like the KACC, funding NGOs like TI-Kenya and cajoling the government into passing procurement bills and adopting ministerial codes of conduct, fully-engaged donors could keep nudging the leadership towards greater accountability. The very fact that ministers had resigned over a corruption scandal – an unprecedented event in Kenyan history – was seized upon as a sign of progress.

'I use a long-term time horizon, which is frustrating for some,' one DfID official in Nairobi told me. 'My attitude is that through a series of gradual changes we are putting in place a system of checks and balances that will almost force Kenya's leadership to improve, despite itself. Not in time for this generation, necessarily, but for the next. This is a bracing time, a very encouraging time, and the overall trend is upwards.' With aid contributing just 4 per cent of the national budget, there was only so much donors could hope to achieve.

Reforming Kenyan society was ultimately the responsibility of the country's citizens, not outsiders, and in the meantime, why should the poorest of the poor – for most aid went on education and health – be made to pay for their leaders' failings? And, by the way, was I so naïve as to think a return to a KANU government would be better for Kenya?

Behind the arguments, behind a politically-correct squeamishness to be heard criticising an African government, loomed an undeniable reality: in the wake of pledges made at Gleneagles to double and triple aid, Western development ministries *needed* to hike disbursement if they were to use the fresh funds coming their way. DfID's development budget for Africa more than quadrupled in the ten years after Labour's 1997 election win. And if Kenya was deemed to fall short of the criteria which made a country a deserving recipient of aid, which African nation would qualify?

But not everyone shared such views, and the debates would be fiercest at the heart of the institution dominating the aid industry. In June 2005, James Wolfensohn had been replaced as president of the World Bank by Paul Wolfowitz, George W. Bush's former deputy secretary of defence, a man who, like Robert McNamara before him, seemed to embrace the war on global poverty as atonement for an American military débâcle in which he had played an instrumental role – in McNamara's case, the Vietnam War, in Wolfowitz's, the invasion of Iraq. Building on the foundation laid by his predecessor, Wolfowitz announced that he was putting the fight against corruption, both within the bank and in recipient countries, at the top of his agenda. Sleaze, he said, was the single biggest obstacle to development. 'It weakens fundamental systems, it distorts markets, and it encourages people to apply their skills and energies in non-productive ways.'

It was time, Wolfowitz told his restive staff – 92 per cent of whom had been opposed to his appointment – the bank moved more decisively against 'irregularities'. He acknowledged he was demanding a step change from an organisation that had once regarded any discussion of the issue as unpardonable cheek, but the move was 'horribly overdue.' The era in which career success was measured by the number of projects approved was over. Managers would in future be rewarded 'as much for saying "no" to a bad loan as for getting a good one out the door'. Hundreds of millions of dollars of scheduled aid to Chad, Congo-Brazzaville, India, Bangladesh, Uzbekistan, Yemen and Argentina were halted as the boss set out to invigorate one of Washington's most complacent institutions.

Within days of announcing a raft of new lending, Colin Bruce in Nairobi was forced to publicly correct the impression of a harmonious relationship between World Bank and Kenyan government. In fact, he told Nairobi's journalists, $265 million of loans to Kenya remained on hold, delayed because of corruption concerns. In a strenuous remarketing effort, the same facts were given a radical new gloss: the World Bank and John Githongo were essentially fighting the same battle. Daniel Kaufmann, who had become convinced corruption's impact on developing economies was massively underestimated, was sent to Nairobi to carry out an independent assessment of the lending programme.

Passing through London at the end of January 2006, Wolfowitz invited John to join him for dinner. Not someone who had ever expected to see eye-to-eye with a Bush hawk, John found himself inwardly cheering the neo-con's proposals to restructure lending programmes to focus on the graft issue. When the World Bank unveiled a special panel to investigate the underperforming internal investigation unit created by Wolfensohn, the Department of Institutional Integrity (INT), John was one of six independent experts recruited. The panel was headed by Paul Volcker, former chairman of the US Federal Reserve, who had also probed the UN's oil-for-food scandal. John had made the transition from domestic whistleblower to internationally recognised expert on graft, with a finger on the pulse of the global aid debate.

The fight against corruption had mushroomed out of all recognition, going from an issue championed by a few activists to an industry whose meetings filled five-star hotels with *per-diem*-claiming suits. Hungry for heroes, this was a sector with a limitless appetite for the rare voices of support coming from Africa, the continent where official graft was seen as doing most damage. Courted by academic bodies, NGOs and multinationals anxious for his endorsement, John was increasingly spending his time in the air, flying to speak at an African Development Bank meeting in Ouagadougou, to Washington for working sessions of the Volcker panel, to Ottawa to plan a Canadian-funded anti-corruption research programme, to Dakar to open offices for the Ford Foundation.

His time in exile had hardened him, bringing to the fore a calculating tendency previously diffused in a young man's general bonhomie. He was less generous with his affections, more self-serving in his relationships. 'I've become much more careful, much more distrustful. What has happened – which I'm slightly uncomfortable with – is that I find myself thinking: "So-and-so wants to meet me? How is it going to be helpful to me?" I never used to think like that. I used to think: "Oh, So-and-so is around, that will be fun."'

There was a ruthlessness in his social dealings now that went beyond the careless Githongoing of the past. He would be his usual engaging, curious self with new acquaintances, people it was useful to add to his contacts list. But old family friends, childhood mates, work acquaintances who had helped him in the past but had nothing fresh to offer, would agree dates, rearrange flights and travel up to Oxford or London, only to discover he had switched off his mobile and gone to ground. Githongoing had become such an established aspect of John's methodology, he no longer felt the need to apologise or justify. If he experienced the odd qualm, the contents of his answering machine and email inbox allayed his conscience. As long as the world kept begging him for interviews, talks and consultancy work – and the requests never seemed to stop – he could afford to hurt a few feelings. Like a shark that has to keep moving to survive, he powered relentlessly ahead, leaving ripples of hurt and dismay behind.

His seven-year relationship with Mary was in its death throes. The sniping of supposed friends – 'When are you going to wake up and realise, girl, that man is gay, gay, gay?' – had not helped, but the breaking off of their engagement was the result of profound character differences and irreconcilable aspirations. In contrast, there had been a rapprochement with his parents, who flew regularly to Britain for medical treatment. There was no need to hide things from them now. Properly relaxing for the first time in years, John succumbed during their extended visits to the sudden adrenalin slump that can afflict the highly-strung, discovering an insatiable appetite for sleep. When he surfaced, the family caught up on lost time, discussing recent events over meals.

A mother's indulgence is something a first son can always count on, but a blessing from Joe Githongo, friend of so many of the men he had defied, the man whose business was surely doomed to suffer for John's actions, was less certain. One afternoon, John sat the old man down with a pair of headphones and played him the tapes. 'It was fascinating. The room was hot, the air conditioning was off and he's a diabetic. But he didn't snooze for a moment. He knows these people much better than I do, he has much greater insights into their thinking. He said, "Listen, John, can't you tell? These people have been talking about you all night."' Once again, John watched someone else tracing his own steep learning curve. 'What hit him as a revelation was the extent to which people he considered old friends betrayed me, and through me, him. He said: "You never told us anything of this. We could see it in your face, but you never told us." Afterwards, he kept saying: "You did the right thing. You stayed too long."' Father and son had made their peace.

So long suppressed, John's indignation had surged to wrathful heights. Ironically, he felt angrier in his second year of estrangement than in his first. 'There's a sense in which my own exile has been me coming to terms with the fact that Kibaki, Wanjui, that entire generation, used me. Because I had a great affection for Kibaki, I've had great difficulty coming to terms with that emotion.' He was increasingly conscious that he had been in an extended state of denial, determined not to hear what his more knowing colleagues told him about the head of state. 'I've been going back over the documentation and you suddenly realise, "Of course, they are telling me." They told me several times, "John, the president was doing this type of thing when he was vice president, when he was minister of finance . . . I mean, what do you think?" And I have had a blind spot. I don't know, I can't describe it.' He had thought he fled Kenya in fear for his life. He had also, he suddenly realised, been running from having to face the truth.

But his aspirations stretched beyond Kenya now. He hoped to push the global aid industry into recognising that its chirpy determination to look on the bright side, its readiness to turn a blind eye to blatant abuse, was doing Africa more harm than good. Increasingly, he found himself at odds with the Geldofs, Bonos and Sachs of this world, and the government departments and multi-lending institutions upon which these celebrity campaigners focused their gaze. There was something almost neo-colonial, he felt, about constantly pushing for more aid for Africa, while failing – as so often happened – to ask searing questions about the quality of the leaderships responsible for disbursal. The wristband-wearing activists who linked hands around Edinburgh in solidarity with the Make Poverty History cause might bask in the glow of moral righteousness, but to John, an unarticulated 'It's Africa, what else can you expect?' lay behind their pitying stance. 'There's a condescending, implicitly racist argument with regard to Africa, which says that "excessive enthusiasm" in the fight against corruption somehow undermines the task of fighting poverty. But corruption, systemic corruption, is the most efficient poverty factor on the continent.'

Such views put John at odds with those he had once regarded as allies. His enemies in Nairobi were convinced he regularly briefed British officials on the innermost workings of the Kibaki government. Even I had assumed that, as soon as his exile became public, Kenya's donors would beat a path to his lodgings, anxious to learn the sordid truth about an administration entrusted with hundreds of millions of dollars of their taxpayers' money. Not exactly. It would be two and a half years before any Western institution asked John for a debriefing on Anglo Leasing, and the request, when it was eventually made, came from the Serious Fraud Office, not from any department responsible for shaping British policy abroad. Kenya's donors had no need to quiz John, for they had understood the scandal's contours even before he had, and had determined to carry on lending regardless.

The tone of John's relations with Britain's development ministry was set during a lunch at a London hotel with Simon Bland, head of Nairobi's DfID office, and Dave Fish, DfID's director for Africa, called shortly before John was due to hold his one and only encounter with Hilary Benn, the then secretary of state for international development. 'It was a very unpleasant meeting,' John remembers. 'The message was: "This is Africa, it's always been corrupt." They were very provocative and sneering, talking through gritted teeth. They kept referring to my "allegations", basically saying: "You've upset our programme." I realised afterwards I'd been talking to two very angry men.' Like the 'Hear No Evil' monkey, hands clamped firmly over their ears, humming furiously to drown out the voices, donors had no interest in what John had to say.

Aware that he was a guest in a foreign land, John did not complain about the British authorities' indifference. But there were others ready to do so in his place. 'It's damnable. Why they won't act, I just don't know,' fretted Edward Clay. 'John is a one-off, someone like him won't come along again. In future, we'll have intimations of corruption, the odd hint from MPs in parliament, but we'll never have a case again in which someone with John's insights and access offers the dope up on a plate.' By failing to take forceful action, Clay believed, DfID and its fellow donors had set the worst possible precedent, not only for Africa, but to the recipients of British aid across the globe. If the donors were not going to make an example of Kenya over Anglo Leasing, it was hard to see when they would ever get tough. 'By their negligence, DfID have shown that they don't wish to see this issue pursued to a kill. And that's unforgivable. The Kibaki government has said, by its actions: "We will reward and reappoint those involved in Grand Corruption." And we've said: "OK." The message we're sending the Kenyan people is that on the whole, we'll always line up with those in power.'

* * *

The Public Accounts Committee's hard-hitting report into Anglo Leasing, the outcome of John's Kenyan High Commission testimony, was tabled in parliament in March 2006,[37] as was the auditor general's equally damning account. The PAC report accepted that Kibaki had known about the Anglo Leasing contracts, recommended investigation of vice president Moody Awori, former finance minister David Mwiraria, justice minister Kiraitu Murungi and civil service head Francis Muthaura, and found attorney general Amos Wako guilty of serious negligence. It called for prosecution of those responsible, recommended termination of pending Anglo Leasing projects and said there should be no payment on projects already started until due diligence had been carried out. But there was a vast gap, in a country where real power rested in the hands of the president, between the recommendations of bolshy MPs and concrete action.

On the ground, the trend was moving in the opposite direction. Kenya's highly politicised legal institutions and judicial system were about to start systematically erasing the anti-corruption work done in the first months of NARC rule. A first step in this process, laden with symbolism, was the clearing in July 2006 by the Constitutional Court of George Saitoti over the Goldenberg affair, quashing a recommendation by the Bosire Commission for possible legal action against the former vice president and finance minister. Lawyers predicted that all other pending Goldenberg-related cases would now fall by the wayside. Watching men like Kamlesh Pattni being questioned on live television had been one of the most astonishing breakthroughs of the early NARC era. The spectacle had sent out the message that, even if it took a decade and a half, the corrupt would ultimately be held to account. Now one of the keystones of the incoming government's clean-up programme had been removed.

Kenya's elite had always shown a readiness to use the legal system to strangle criticism. John could see the process at work in his own life. In Kenya he was being sued for libel by Chris Murungaru, former internal security chief, while the shadowy businessman Anura Perera, the man who had bought his father's debt, and Perera's lawyer were both pursuing him separately in the UK, the former hiring the high-profile libel lawyers Carter-Ruck. In the old days, the various suits would have kept John worrying into the small hours. But he had toughened up, actively relishing the twists and turns of this high-stakes chess game.

The Mount Kenya Mafia, he realised, was recovering its equilibrium after a momentary wrong-footing. The shelving of the Goldenberg scandal, he became convinced, heralded an imminent mothballing of investigations into Anglo Leasing. 'There's a massive rollback going on.' It was time, he had decided, to give the spinning top that was Kenyan public opinion another flick. 'We're going to go for Round Two. You know, we never finished the first round, we fired a first blast and waited. They think it's over. So I need to make it clear it's not.'

This time, he knew, it would be harder to win a hearing. Relations with his former employer, the Nation Media Group, had deteriorated, with distrust building on both sides. 'John has been too slow in releasing his story. We were begging him for more,' managing editor Joe Odindo complained to me that August. 'You have to strike while there is an appetite. You cannot keep hitting people time and time again with these things. People just get tired of it, and so do the journalists. They begin to say to themselves: "Hey, we thought we were releasing a story, not forming part of a campaign."' John's careful husbandry of his material had left editors feeling used, while he suspected that behind-the-scenes pressure from NARC, leaning on its friends on the board, explained the newspaper's tangible loss of appetite.

His first targets had been the ministers and civil servants who had negotiated and signed the Anglo Leasing contracts. The men and institutions notionally responsible for delivering justice, so compromised that they had become part of Kenya's sleaze problem, rather than its cure, would be next. This time his target would be his supposed former brother-in-arms, the venerable head of the Kenya Anti-Corruption Commission. In drafting his dossier, John had avoided all criticism of Justice Ringera. He had done so partly for tactical reasons – fighting on more than one front is never a wise strategy. But there had also been an emotional element to his self-censorship. Despite everything, John had harboured a grudging affection for Ringera, a man he saw as trapped in a near-impossible job. He had hoped that within the limits of what State House allowed, the judge might find some way of reining in Anglo Leasing's perpetrators. That belief had evaporated, and with it any magnanimity.

The building housing the KACC is not particularly ancient, but it has aged at turbo speed. Located on the site of a former nightclub, it was once the site of Trade Bank, one of the most notorious of the 'political' banks set up in the 1980s by Moi's cronies, eventually destroyed by illicit loans. The premises became available when Trade Bank's Asian founders fled the country, leaving depositors seduced by their 'No Hassle' motto to mourn their folly. That history prompts a certain mirth. 'I mean, you take the spot where one of Kenya's biggest scams took place and call it "Integrity Centre"?' guffawed a Kenyan Asian I met at a party. 'Are they taking the piss, or what?'

The building's architects clearly saw this as a cutting-edge project, its glass front, tubular design and shiny metallic panels intended to transmit a message of brash efficiency. But that was before Nairobi's rains got to work. The bronze-coloured panels are now rusting along the edges. Long orange drips stain the once-shiny surfaces. As for the centre's blue neon sign, some of the connections have gone, leaving parts of the giant lettering invisible at night. Whenever I return to Nairobi I like to drive past it, to see what it is telling passers-by. 'INTEGRITY CENTRE' it once proclaimed. Nowadays, it reads: 'INT⁻GFI Y CENT¬E'. The garbled message seems appropriate for an organisation which has lost its way: an addled sign for a befuddled institution.

Theoretically, Justice Aaron Ringera enjoyed an enviable level of independence. Blessed with security of tenure, running an office generously supported by foreign donors, banking what was – thanks to John's energetic advocacy – a staggeringly generous salary, he could be held accountable only by parliament. His critics said none of this mattered when weighed against a key affiliation, one betrayed by a linguistic quirk. Ringera has problems differentiating between 'r's and 'l's. 'Lobbying' comes out 'robbying', 'present', 'plesent'. It is a foible prevalent among the Mount Kenya community. Like former justice minister Kiraitu Murungi, with whom, in one of Kenya's classic Venn circles, he was once a partner in a law firm, Ringera comes from Meru. The day we met he boasted that he had never taken a telephone call from the presidency during his time in office. But a former colleague told me he never needed to, going round in person to State House to receive instructions. 'He would call staff together and say: "The message from the president is A, B, C and D."' So perhaps it is appropriate that the interior of the KACC feels exactly like a government ministry, with its dark wood furniture, stiff flower arrangements, padded beige walls and obligatory Thermos flask.

With his corona of white curls, the judge looks like an ageing cherub. He is charming, confiding, and had not a malicious word to say about his former ally – 'John was my friend, he *is* my friend' – merely softly chiding the younger man for failing to grasp the limits of his role. 'When you are an adviser you are not the chief executive. Your advice may be taken, or ignored. John expected to be more important than he was. In that way he was naïve.' Behind the cuddly façade, one caught a glimpse of a complex, sophisticated individual well aware of the contradictions of his own position. And behind that, one sensed a colossal laziness, the complacency of a man who was exactly where he wanted to be, in career and monetary terms, and intended to stay there as long as possible.

The first hint the Kenyan public got of Ringera's less than helpful role came in September 2006, when Martha Karua, the new minister for justice, announced she would soon release a list of those responsible for Anglo Leasing. Blaming all the dodgy contracts on the previous regime, she said investigations had been delayed by John Githongo's failure to sign a statement authenticating his KACC testimony in London. This provocation could not go ignored. In a three-page statement to the press, John now revealed exactly what had taken place in the Kenyan High Commission in Portland Place the previous March.

Understandably wary, John had insisted his evidence be recorded. Yet at the end of two days of testimony – long enough, one would have thought, for any malfunction to be detected and rectified – Ringera informed John the recording equipment had failed and his words were, sadly, inaudible. The taper had not been taped, so a KACC-drafted summary would have to take the place of the audio transcript. After delivering that bombshell, Ringera waxed astonishingly candid about the charade being staged for public consumption. There would be no Anglo Leasing prosecutions until after the 2007 elections, if ever, he said. When John's lawyer asked when his client could return to Kenya to give evidence, Ringera said: 'No, no, I wouldn't advise that.' John should not underestimate the pain he had caused certain people. It would not be safe. Once again, the head of Kenya's anti-corruption body was doing his best to halt investigation.

John's account would almost certainly have been dismissed as a fabrication by the establishment, had it not been for one man's surprise intervention. A former lead KACC investigator who had accompanied Ringera to London went public to confirm the details of the conversation. 'I was amazed by what Ringera said. There was no indication whatsoever the equipment wasn't working. You don't go for days without once going through what you have recorded,' he told me.

The man concerned has featured before in these pages. It was Hussein Were, the quantity surveyor who was once squeezed out of his former workplace by ethnic favouritism. He had joined the KACC, only to discover a new form of ethnic discrimination, this time pitched in favour of the Mount Kenya community. Having travelled with a colleague to Switzerland, France and Britain on the KACC's behalf, attempting to establish a paper trail linking Anglo Leasing's companies with bank accounts and principals in Kenya, he had seen their recommendations for further action ignored. 'We were convinced that if we took those steps we were definitely going to answer the question at the heart of this affair: "Who really is Anglo Leasing? Who are the principals, who are the beneficiaries?"' Were had become convinced no Anglo Leasing prosecution would ever take place as long as the judge, loyal defender of his ethnic community, headed the KACC. 'It is not by accident that Ringera is still there.' In daring to publicly defend John Githongo's testimony in the Kenyan press, Were had once again demonstrated that there were things that mattered more to him than a quiet life.*

If Ringera was embarrassed by these revelations, he did not let it show. Ignoring opposition calls for his resignation – he had tenure, after all – he sent a handful of supposedly 'watertight' files to attorney general Amos Wako, recommending a dozen Anglo Leasing prosecutions. The attorney general promptly sent them back, claiming they were too incomplete to be acted upon. Conveniently, each man could blame the other for the failure to proceed, leaving the public baffled as to who was really responsible. The purely cosmetic nature of all this seeming activity became clear on 6 November 2006, when Ringera exonerated former justice and finance ministers David Mwiraria and Kiraitu Murungi of a preliminary charge of obstructing justice. John's testimony, ruled Ringera, was of no legal significance because he had not held the status of official investigator. It was an infantile justification for inaction – a crime is a crime, whatever the supposed status of the person uncovering it – but it would be grabbed hold of and repeated incessantly in future by the Mount Kenya elite. Kibaki promptly reappointed Kiraitu and George Saitoti to his cabinet. 'I will never speak about Anglo Leasing any more because is it no longer an issue,' declared a delighted Kiraitu. 'It is dead.' After less than a year in the cold, two of the three ministers who resigned after the leaking of John's dossier were back.

'In war, it is not good practice to discharge all your ammunition in the first battle,' John had warned his adversaries. He now demonstrated what he meant with the release of a second tape, this time posted directly onto the internet. On the tape, recorded in June 2004

* Hussein Were was sacked by the KACC in June 2006, accused by his superiors of leaking stories to the Kenyan media. He denies this, and believes he once again fell foul of an ethnically motivated purge.

in Kiraitu's presence, an unhappy, audibly jittery Mwiraria frets: 'This thing, this thing . . . If we are not careful, will come down with our government.' 'Drop this matter,' he begs John. 'I will get to the root of the matter, I will find out who it is in my own way.' The recording stayed online long enough for it to be downloaded and relayed by every Kenyan media outlet, finally crashing when a saboteur posted Kenya's economic recovery strategy repeatedly on the website. John had made his point. This was what the men exonerated by the KACC director had got up to.

When I quizzed him about this entire, mortifying episode, Ringera took refuge in legal niceties, as any skilled attorney knows how to do. There was a difference, he explained, between an illegal and an immoral act, and in the case of Anglo Leasing, many of the things that shocked John – the justice minister's blackmail attempt, for example – fell into the latter category. 'You could say morally it was terrible, but according to the Penal Code, no offence was committed.' John's evidence in itself was mere hearsay, and the law did not stop a president appointing ministers who had not been charged with a crime. 'In the eyes of the law they are whiter than white.' As for Ringera's habit of passing on death threats, he was baffled by John's interpretation of these exchanges. 'I was telling John that as a friend. I don't know how he could have seen that as threatening. We were very close.'

Confronted by such brass face, John's salvoes fizzled and died. Kenyan commentators labelled his revelations 'the second Githongo dossier', but it was obvious to all that their impact did not measure up to the first. With a series of KACC absolutions and ministerial reinstatements, the Kibaki administration had cynically set out to test domestic and international opinion, and had discovered that it could reverse the concessions of the previous year without serious repercussion. John was merely scoring points now, rather than changing the shape of Kenyan politics.

When I met John's brother Mugo in Nairobi, he acknowledged the shift. Since we'd last seen one another, he had become inured to the intelligence men who followed him around town, sitting within eavesdropping distance when he met friends in bars and restaurants.

If he had adjusted to the post-dossier Kenya, so, he reckoned sadly, had others. 'A lot of the guys who were engaged when John released his dossier are missing in action this time around.'

On the international stage, too, the air was hissing out of the global anti-corruption drive. By late 2006, Paul Wolfowitz's position at the World Bank was looking increasingly beleaguered. Described by one former colleague as a man who 'couldn't run a two-car funeral', Wolfowitz had always been respected for his intellect but regarded as an incompetent manager, and his stint at the World Bank had exposed this administrative ineptness for all to see. He had ruffled feathers by surrounding himself with aides recruited from the Bush administration, abrasive young Americans who were allowed to overrule, bypass and humiliate non-US bank directors with decades of experience. Incapable of delegating, secretive by nature, Wolfowitz took decisions alone in his office, only to discover on emerging that they violated the procedures of this most rule-obsessed of organisations.

As he alienated potential allies, his anti-corruption strategy took a hammering, with Britain's DfID, champion of higher aid allocations, leading the way, and non-governmental organisations, which had always viewed Wolfowitz as a White House plant assigned to politicise a supposedly neutral international institution, piling in enthusiastically. Wolfowitz, argued the government ministers whose contributions kept the World Bank in business, was making the poor pay for the sins of the elite. The fact that Colin Bruce toured Western capitals, lobbying the same men and women to unfreeze pending loans to Kenya and approve fresh ones, was a measure of the mutinous mood. It was an initiative, said bank insiders, Bruce would never have dared to take without the blessing of influential colleagues and key ambassadors.

There was a strong element in the gathering revolt of score-settling for American high-handedness in its War on Terror – 'payback-by-proxy', as the *Washington Post* termed it, for the Iraq war. That was understandable, but in the process of taking their revenge,

Wolfowitz's critics risked throwing the baby – an overdue examination of the World Bank's complicity in corruption – out with the bathwater. John had no doubts on which side his sympathies fell. 'There's a struggle for the soul of the World Bank taking place, and sadly that fight is being engaged over the small bit of paper called the anti-corruption strategy. The *nomenklatura* is pouring cold water over everything Wolfowitz suggests, deliberately misinterpreting his anti-corruption paper in order to kill it off.' By the time Wolfowitz's cherished Governance and Anti-Corruption (GAC) Framework was presented to the bank's Development Committee at its annual meeting in Singapore in September 2006, it had, in the view of its initial supporters, already been diluted by the institution's old guard to the point where it had become a collection of woolly aspirations rather than a pragmatic plan of action. Hilary Benn, DfID's international development secretary, took the opportunity to stick the stiletto in a little further, announcing that Britain would withhold £50 million in World Bank funding unless onerous conditions on lending to poor countries were removed.

The year's end saw two massive setbacks to John's fight against corruption. On 14 December 2006, Britain's attorney general, Lord Goldsmith, suddenly announced that the Serious Fraud Office had halted a long-running probe into allegations that BAE Systems, the British defence contractor, had paid out millions of pounds in bribes to Saudi dignitaries in exchange for military contracts. The suspension, which came just as investigators were on the brink of accessing Swiss bank accounts, was justified on the unconvincing grounds of 'national security': continuing, prime minister Tony Blair said, would have damaged Britain's relationship with a key ally in the War on Terror. But there was no disguising the ugly fact that a country which had signed the OECD anti-bribery convention had agreed to investigate no further when a high-profile company came under pressure from an important foreign customer. In one fell swoop, Britain's outspoken stance on corruption was made to look like so much hypocritical posturing, its ability to criticise sleazy foreign states drastically undermined. As the Liberal Democrat MP Norman Lamb put it: 'How on earth can we lecture the developing world on good governance when we interfere with and block a criminal investigation in this way?' Contemptuous laughter resounded across the African continent. 'What the BAE case shows,' said John, 'is that Edward Clay didn't represent a change in British policy. He was an anomaly.'

The other blow was geostrategic. Taking advantage of the world's inattention during the holiday period, on Christmas Eve Ethiopia sent its army rolling into neighbouring Somalia to deal with the burgeoning Islamic Courts movement there. The United States, Ethiopia's closest ally, seized the opportunity to launch air strikes on suspected Al Qaeda operatives among the scattered Somalian forces fleeing south. Kenya played a helpful supportive role in the operation by closing its border, moving its army up to the frontier, and later flying suspected members of the Islamic Courts who fled onto its territory back to Somalia and Ethiopia for interrogation. When put to the test on an issue close to the Bush administration's heart, Kibaki's government had more than delivered. It was not hard to guess which, in future, was going to matter more to Washington – the fight against corruption, or the loyalty of a strategically situated ally in the War on Terror. However exasperated its ambassador on the ground might feel, Washington was unlikely to crack the whip in its future dealings with Kenya, and Kibaki's team knew it.

By early 2007, the World Bank was in a state of near open mutiny, with staff who wanted Wolfowitz gone openly sporting blue 'good governance' ribbons to work. Then, in April, the boss's enemies were suddenly handed the excuse they needed. Wolfowitz, it emerged, had personally approved an overly generous severance package for his girlfriend, previously also employed at the World Bank. Shaha Ali Riza's transfer, ironically enough, had been motivated by the laudable desire to avoid any conflict of interest, and what Wolfowitz had done paled into insignificance when compared to what government ministers in the World Bank's recipient countries regularly got up to. But a man leading a global war on corruption had to be whiter than white. In May 2007, Wolfowitz accepted the inevitable and resigned.

The Volcker panel on which John sat delivered its findings four months later, recommending, as part of a proposed overhaul, the elevation of the internal investigation unit's head to the rank of vice president. But the wind had been knocked out of the World Bank's anti-graft campaign. By December 2007, the bank's representative Colin Bruce could proudly point to a Kenyan portfolio of sixteen active projects, worth $1 billion, with another $260 million due to Kenya from its share in three regional projects. Truly, this country director – now living in a compound whose perimeter wall and entrance gates had enjoyed a presidential upgrade at taxpayers' expense – had pushed the money out of the door.

16

A Plaza Paradise

> 'People in this country are like meat for hyenas. The only question is which hyena do you prefer to be eaten by: Hyena Raila, Hyena Kibaki or Hyena Musyoka? Whichever it is, it's still a hyena coming to eat you.'
>
> Nairobi kiosk-owner

In September 2007, huge billboards went up across Nairobi, looming over its busy roundabouts and honking junctions. 'Hummer is here,' ran the ominous message. 'Maybe you felt the tremors.'

And so we had. General Motors' Kenyan launch of the Hummer H3, one of the most macho 4x4s on the market, was one of those small events which combine with a host of other details to send out a signal about where a country is heading. It came as no surprise that Raila Odinga, always a flamboyant performer, beat everyone else to it by importing one of the gas-guzzling behemoths, dubbed 'The King of Bling' by motoring fans, months ahead of the launch. But what took the breath away was the announcement by General Motors' local director that he expected to take a hundred orders for the car, priced at a sobering six million shillings (£47,000), in the first year. Did one hundred people really have that kind of cash in a land where the average citizen earned just $460 a year? Had J.M. Kariuki, the left-leaning populist who warned Kenyatta of the dangers of creating a Kenya of 'ten millionaires and ten million beggars' shortly before his assassination, only got his figures slightly wrong?

Certainly, as the presidential elections beckoned, the nation seemed in the grip of an urban spending spree whose brash ostentation was a world away from the dourness of the belt-tightening Moi years. The stock market was booming, the shilling strong, agriculture enjoying a recovery, tourist numbers hitting new records, tax revenue – boosted by the newly efficient Kenya Revenue Authority – higher than it had ever been. Was it possible the old boast of Central Province – that only the Kikuyu could be trusted to run the country's economy – had something to it after all?

Once Nairobi had been known as the Garden City, where your breakfast could be stolen by light-fingered vervet monkeys sneaking in through an open kitchen window. Now it seemed a City in Permanent Construction, its birdsong drowned out by the sound of constant drilling, greenery thinning amid the high-pitched whine of electric saws. The self-confidence and technological nous of the Kenyan diaspora, which had brought its savings back and was investing at home, were reshaping the city. Cynics pointed out that construction had always been the easiest way of laundering illicitly acquired funds, but the sense of gathering momentum was beguiling. Across the capital, empty plots of land evaporated with the speed of puddles in the African sun. No street corner was now complete without a new apartment block in the local blue-grey Nairobi stone, and at the end of many of those streets the traditional two-storey shopping centre, with rows of small metal-grilled Asian shops, was dwarfed by a giant plaza offering seven-day shopping, twenty-four-hour service, beauty parlour, cinema, ATM banking, internet access and a branch of the Java café chain, *the* venue of choice for the city's latte-drinking, BlackBerry-wielding, laptop-addicted young professionals. You could measure prosperity levels in a new phenomenon: the Nairobi traffic jam. Once an exclusively rush-hour feature, it now seemed to last all day. Why, these days Nairobi even boasted an ice-rink – one of only three in Africa – where squealing Kenyan boys and girls tottered across the ice and thumped against the wooden barriers.

If Kibaki's government wasn't much good at delivering roads or affordable housing, it showed an impressive enthusiasm for the kind of purely cosmetic makeover calculated to warm the heart of the most pursed-lipped of bourgeois housewives. Nairobi had become one of the few capitals in the world to outlaw smoking outdoors, a rule enforced with a six-month jail term. Another ban was briefly slapped on plastic bags – *so* unsightly – and another on livestock inside the city centre: those herds of cattle and goats being driven along main roads by Maasai herdsmen were nothing more than an embarrassment, a reminder of Kenyan society's humble roots. Thousands of the kiosks clustered on road verges – the cardboard-and-corrugated-iron shacks in which the *wananchi* ate their lunches, drank their sodas, gossiped and slept – were ruthlessly bulldozed and replaced with neat flowerbeds and tended lawns. Determined to broadcast the slickest of images to the world, the administration found such evidence of blatant poverty mortifying.

Kenya's favourite reading matter offered a telling insight into this strand of the national psyche. In Nairobi bookshops, three times as much space was dedicated to Western self-help books as to African politics or history. *Feel the Fear and Do it Anyway*, *How You Can Get Richer Quicker*, *The 7 Habits of Highly Effective Families*, *The Magic of Thinking Big*, *Why We Want You to be Rich*. Nothing, explained a salesman at Text Book Centre in the Sarit Centre shopping plaza, sold faster than these motivational books, which enjoyed a display all to themselves. 'This shelf makes my day,' he enthused. 'People are looking for ways to create a positive environment. If I think negative things, I portray negative things.' He was, he admitted, a keen consumer of his own wares. 'These books give me confidence and a purpose. It's part of a different attitude in Kenya. Essentially,' he explained, 'it's all about getting rich.'

'My favourite word is "Aspirational",' declared Aly Khan Satchu, a Kenyan Asian who returned from the City of London to launch a stock market analysis website in one of those very plazas. His website was baptised www.rich.co.ke and promised: 'Strike it rich.' The adjective he'd picked was virtually a national theme tune. 'How do you see your average reader?' I'd ask the young editors who launched a spray of new metropolitan newspapers, business weeklies and lifestyle magazines that year. 'Very aspirational,' came the uniform answer. 'Graduates with job prospects and disposable income.' The same word cropped up in the excuses proffered by Kenyan companies refusing to endorse NGO community projects in the slums that everyone tried so hard to forget: 'Sorry, not an aspirational enough audience.'

Were there really enough 'aspirational' people in Kenya, I wondered, to keep all these malls busy? Or, to put it slightly differently, weren't all these thrusting entrepreneurs in danger of forgetting that the most genuinely 'aspirational' segment of Kenyan society was not in fact its small middle class, but the millions of exasperated inhabitants of Korogocho, Dandora, Mathare Valley and Kibera? And what would happen when those Kenyans finally registered that while a tiny elite was 'eating' as never before, their own, more modest aspirations were doomed to go forever ignored?

Perhaps the most worrying element exposed by economic surveys was the extent to which, once you set aside the cosmopolitan cities, the growing divide between rich and poor took geographical – in other words, ethnic – form. A Kikuyu inhabitant of Nyeri, just north of Kibaki's constituency, could expect to live 23.4 years longer, on average, than his Luo counterpart in Raila's home town of Kisumu. If 46 per cent of the population in Central Province had only limited access to a qualified doctor, the problem was nearly twice as bad – 88 per cent – in remote North-Eastern Province. Adult illiteracy, just 16.7 per cent in largely Kikuyu Thika, was 78.1 per cent in Bomet, a heavily Kalenjin Rift Valley town.[38] And so it went on.

The regime's critics noted that many of the superficial features of the former era had crept back, so quietly they almost went unnoticed. Kibaki had promised to keep his name and image off Kenya's currency, institutions and roads. Now his official portrait hung on the wall of almost every office, just as those of Moi and Kenyatta had, and his features were stamped on the new forty-shilling coin. The man who, while in opposition, had promised not to waste Nairobi residents' time by blocking the city centre with official motorcades, now regularly paralysed traffic for hours at a time, prompting one outraged letter-writer to the *Daily Nation* to suggest he try using a helicopter instead. Trivial disappointments, perhaps, but behind them lurked massive betrayals: the failure to reform the constitution, the failure to devolve power, the failure to appoint a prime minister. However noble the Kibaki government's intentions had been at the outset, yesterday's radicals had by 2007 turned into the steeliest of reactionaries, propping up a system they once abhorred. And *kitu kidogo* had made a comeback, with ordinary Kenyans – a Transparency International survey revealed – encountering corruption in more than half their dealings with officialdom.

One of the lessons of the previous five years seemed to be that when the spirit wasn't willing, it really didn't matter how many worthy new institutions, appointments or laws a government unveiled: the status quo remained unchanged. Witness the fate of the Public Officer Ethics Act, part of the raft of anti-corruption laws proudly announced on the lawn of State House in the wake of Kibaki's inauguration. Hailed as landmark legislation, it required government employees to declare their wealth, in the hope that this would prevent them using their positions to line their pockets. Yet the Act was rendered ridiculous from the outset. While countries like Tanzania and Uganda only required employees in key positions to fill in the declarations, Kenya made it compulsory for all 660,000 civil servants, whether drivers, messenger boys or volunteers, and extended the requirement to their spouses and children. The information provided was confidential, was not computerised, and since the declarations were to be kept for thirty years, storage space soon became a problem.

Having created a tsunami of information, the Act failed to specify how it should be analysed or what action should be taken if wrongdoing emerged. By September 2007 there had not been a single case of a public servant being prosecuted or even fined under the legislation. Who was to blame for this exercise in futility? According to Erastus Rweria, head of the Efficiency Monitoring Unit in the Office of the President, a group of parliamentarians who had served in the Moi government had deliberately neutered the Act. 'The MPs' aim was to make it unuseable. Most of them are in parliament to protect what they grabbed.' No doubt. But it was difficult to imagine, in light of Anglo Leasing, that government ministers of Kiraitu, Mwiraria and Murungaru's ilk would be too distressed by this act of sabotage.

Nothing better illustrated Kenyan society's acceptance of its own glaring faults than the rehabilitation of Kamlesh Pattni, architect of the Goldenberg scandal. By 2006, the man who had nearly destroyed Kenya's economy had renounced his Hindu faith, embraced Christianity and been reborn as 'Brother Paul', preaching from a hall inside a casino complex. When journalist Kwamchetsi Makokha was assigned to interview the sleek former jailbird on live television, he was taken aback by what followed. 'All these young people who had been manning the lights and cameras suddenly rushed up and mobbed Pattni like groupies. They were all excited, asking for his autograph, one even held out his sleeve for Pattni to sign.' Salim Lone, spokesman for Raila Odinga's Orange Democratic Movement (ODM) opposition party, was attending a funeral for a group of MPs killed in a plane crash in north-eastern Kenya in April when he heard a storm of applause. 'I assumed some celebrity had arrived. But no, it was Pattni. They were applauding him like some kind of hero.' Far from earning society's opprobrium, one of Kenya's most outrageous conmen had acquired the glamorous aura of a rock star. He had done what so many dreamed about but did not dare attempt.

Perhaps the ultimate act of cynicism came in August 2007 when, to the astonishment of his own Kalenjin kinsmen, former president Daniel arap Moi suddenly announced his support for Kibaki's re-election bid. The probable explanation for this baffling move came shortly after, when a 2004 version of the report Kroll had been compiling into Goldenberg's missing millions was leaked to the press. Kroll had abandoned the project when the Kibaki government stopped paying its fees, but the draft still shed devastating light on the systematic looting conducted by Moi's family and friends. The former president's sons Philip and Gideon were reported to be worth £384 million and £550 million respectively, with the assets held in an array of international real estate, bank accounts and shell companies. The implication of the leak was clear: Moi had given his political endorsement in return for a promise that he would remain free to enjoy his stolen assets, whose location the government had known for years but done nothing to recapture. A few months later, a section of a bill was quietly approved granting amnesty to anyone who confessed to grand corruption and offered their illicitly-acquired assets in amends. Section 56B of the Anti-Corruption and Economic Crimes Act miraculously appeared on the statute books despite having earlier been deleted by parliament. 'Help us remain at the trough, and we'll let you continue eating,' one administration had told its predecessor.

The ruling class seemed locked in a mood of amoral pragmatism. If the economy was thriving, what did the unsavoury realities denounced by anti-corruption campaigners really matter? To use a cliché beloved of economists, the rising tide of prosperity would surely end up floating every citizen's boat, and a little sleaze was no more than scum on the water – unsightly, but not doing any real harm. When challenged about graft in the media, government ministers always responded by pointing to growth rates, apparently unable to grasp that the issues – economic prosperity and individual theft – were, in fact, distinct. 'We can afford to be corrupt, given the kind of growth we're delivering,' was the implicit boast.

The international community seemed of like mind. No one would have guessed from its behaviour that Kibaki's government deserved anything other than unqualified support. In May 2007, the United Nations actually awarded Kenya its annual Public Service Award. No April Fool's, this: a government whose key ministers and top civil servants had conspired to steal up to $750 million in public funds was commended for 'improving transparency, accountability and responsiveness in the public sector', with Kenya beating states like Singapore, Austria, India and Australia. In September came another prize: the World Bank ranked Kenya as one of the world's top ten reformers when it came to ease of doing business. And that same month, a beaming Amos Kimunya, the man who had gone to Oxford to beg for John Githongo's silence, signed a joint aid strategy with seventeen foreign donors, including Britain, the United States, Japan, the World

Bank, the EU and the UN, giving the Kenyan Treasury a greater say in how aid funds were spent. Aimed, the signatories said, at 'improving the effectiveness of development aid', the agreement also allowed the government to closely monitor aid being channelled to civil society. Anglo Leasing might never have existed. And the same went for John.

It was a Saturday afternoon at the Karen Country Club, that time of day when the club's golfers stroll in off the greens and settle on the leather banquettes in the wood-lined bar to play the role so relished by the Kikuyu elite: that of African English country squire.

Across the rolling lawns, shaded by the kind of giant fig tree the golfers' forefathers would once have regarded as sacred, the caddies padded away, their job done. Above the tennis courts, the giant knuckles of the Ngong Hills were delicately outlined in indigo. Wallets were opened to allow private bets to be settled, chits were signed and beers downed, producing a chorus of low, deep belly burps. Gradually, the conversation got round to John Githongo, who many of the golfers – prosperous businessmen, lawyers and doctors in their fifties, sixties and seventies – had known as a youngster. 'Too young for the job,' said one. 'Amateurish, hugely naïve,' chimed his businessman neighbour. 'It takes time to tackle corruption. John expected things to happen from one day to the next, and so do the donors. It's very immature.' 'Ah, but it was worse than that,' said an older friend, a retired bureaucrat. 'To take confidential information that you came across in your job as a government employee and pass it on to foreign governments, now that's an act of gross betrayal. I was a civil servant under both Kenyatta and Moi, and that would have been a hanging offence.' Voices were rising now as they warmed to their theme. I began to recognise a particular tone to the conversation. These statements had the vehemence of arguments rehearsed so many times they had become a kind of rote-recital, exercises in group affirmation rather than real debate. 'The man was a spy,' said a lawyer, 'a spy who was recruited' – his eyes widened, his finger jabbed the air for emphasis – 'yes, *recruited*, by the British embassy.' There was a chorus of nods and murmurs. 'Yes, yes, he was a spy.'

Now the conversation shifted a gear. John's betrayal, they felt, could not be separated from its wider context, a long-held colonial-era grudge. The fact was, said the businessman, looking directly at me, that the Brits had been happy to do business with Moi, but they had always had it in for Kibaki. Why? Simple: because he was a Kikuyu. I shook my head, aware I was beginning to flush with irritation. But the club member who had invited me agreed. 'The Kikuyu were the first Africans to fight colonialism on the continent. After the Kikuyu came the Algerians against the French. But we were the first. And the fact is that the British have never, ever forgiven the Kikuyu for fighting them, not till this very day, and that is why they are out to destroy a Kikuyu presidency.' The civil servant joined in, with a certain pride: 'We were the Mugabes of the day, the bad boys.' 'They hate us,' said the businessman. 'They really hate us.' In a mood of belligerent victimhood, the beers were finished off and the golfers headed off to their SUVs, gleaming in the car park.

I've had that conversation many times now, in different locations and with different individuals, but always Kikuyu men of a certain age and class. I have tried to persuade them that the obsession driving – if not warping – British policy in East Africa these days is not repressing the Kikuyu but delivering on the Millennium Development Goals; that my country's baby-faced foreign and development ministers are probably more familiar with the Arctic Monkeys' back catalogue than with the history of the Mau Mau, and that to these New Labourites, products of the 1960s, the colonial rationale feels about as familiar and appetising as hoop skirts and spats.* I have never succeeded, even though the statistics bear me out, showing, for example, that British aid to Kenya under a 'hated' Kikuyu president is greater than it was under the Kalenjin Moi, and that Kenyan imports

* New Labour's arm's-length stance towards Britain's imperial history was captured by Clare Short in 1997 when, as newly appointed international development secretary, she wrote to Zimbabwe's Robert Mugabe, who had been pressing for a greater British contribution to the cost of land distribution. 'We are a new government from diverse backgrounds without links to former colonial interests,' she said.

from British firms – supposedly smarting at being boxed out of the Kibaki economy – have actually risen by over a third in the same period.[39] These Kikuyu gentlemen simply refuse to believe that the horrors of Emergency, which loom so large in their own minds, could be so easily forgotten by their former imperial master, just one of a succession of historic misadventures to be mentally filed away. A $750-million procurement scam has been miraculously transformed into a colonial vendetta.

The irony bites bone-deep. For very few of these aggrieved golfers came from families whose young men played deadly hide-and-seek with British soldiers in the dank forests of the Aberdares. Today's Kikuyu elite traces its roots to the other side: to the Home Guard loyalists who helped the British crush Mau Mau and were handsomely rewarded for their collaboration when the whites pulled out. In fact one of the men fulminating over his beer that day was named, much to his irritation, as a prominent Home Guard member assigned by the British to 'soften up' recalcitrant Mau Mau detainees in the internment camps in a recent book by an American academic. In positing John Githongo as a pawn in some dastardly post-colonial plot, the Kikuyu landed gentry had cheekily rewritten their own role in the past.

But perhaps the most revealing aspect of these conversations is what is always omitted. No one, in these exchanges, even bothers to claim the Anglo Leasing deals were honest transactions, or that the ministers involved weren't on the take. No Kikuyu, especially Kikuyu of this social *milieu*, is naïve enough to believe that. Beneath their belligerence lies the same tacit acceptance conveyed by the refusal of every politician named in connection with Anglo Leasing to put his side of the story when I asked for comment. The issue, for these men, is not guilt or innocence, but loyalty. They cannot forgive the fact that the man who exposed a Kikuyu administration to public ridicule came from within. As the businessman told me that day in the Karen Country Club: 'We have a saying in Gikuyu: "No matter how faded or shabby she is, she's still my mother." John should have felt that.'

* * *

In choosing to become the 'Anti-Kikuyu Kikuyu', as he was dubbed on many websites, John had expected to be reviled by his kinsmen, especially those of this age and class. 'I will not be able to travel in certain parts of the country at night,' he'd predicted when drafting his dossier. 'If my car breaks down in Nyeri and people realise who is in it, I'll be in for it.' What he had not anticipated was the bored disaffection of his own peer group. When I telephoned Lisa Karanja, who had taken over the directorship of TI-Kenya, to arrange a get-together, she said she'd ask another member of John's old anti-corruption unit along. But when we met, she was on her own. Her former colleague had bailed out. 'You know, we've been talking about him for four years and a lot of people are all Johned out,' she explained. 'People feel a bit fed up. They feel he did some very good things, but that there's more to Kenya than this.'

She had been surprised, she said, by the sour reactions of younger Kenyans in civil society – and not just Kikuyu – when John's name surfaced. 'When he was first appointed there was a feeling of "Oh, you're just going to serve as a fig leaf." Then it became hero-worship. Then, with the tape leakings, there was a return to hostility again.' John had become indelibly associated with an interfering donor community which, in African eyes, always pinned its hopes on totemic individuals, ignoring the steady work done by less charismatic people and organisations. 'It's a form of national pride, and in John's regard it translates into a feeling of "You're over there in the UK being a prima donna, but you don't have the guts to come and do this stuff here."'

Even those immune to this mixture of chauvinism and professional envy took issue with John's tactics. Rationing his leaks had certainly wrong-footed his enemies, but he had lost the sympathy of his audience in the process. A PR guru of the Max Clifford variety could have told him that winning public affection requires more than sensational facts and compelling proof, it demands a touchy-feely instinct for how the masses – emotionally fickle, sometimes vengeful – will respond. Absence of spontaneity, the obsessive preparation on which this cerebral operator prided himself, had its price.

John came across as too clever by half, a man playing a game rather than acting from the heart. One newspaper later commented that he had leaked his dossier 'like a man chewing groundnuts from his pockets'. A Kikuyu community worker in Mathare slum summed it up. 'At the start, people said, "This is great." But now he's lost a lot of credibility. People say, "Why doesn't he give us the whole story? It's like a movie: you don't want it in bits. And he keeps giving it out in bits."' If the public wanted the catharsis of a three-act blockbuster, John had offered instead a maddening 'Tune in for next week's instalment.'

As Kenya braced itself for elections, the Anglo Leasing affair looked increasingly irrelevant. John remained an avid follower of Kenyan news, religiously scouring newspaper websites each morning, and being briefed by his network of sources. Over his mobiles came offers of key positions in a future ODM cabinet. 'Come back. Your country needs you,' was the message. He vacillated, packing his things in boxes, shredding his papers, mentally saying goodbye to Oxford. Then he paused, unable to take the final, irrevocable step.

He did not share the donors' faith in the future, noting instead the vast crowds turning out for ODM rallies in the Luo and Kalenjin strongholds of Kisumu and Eldoret, voters who, it was safe to assume, could not afford the Hummer H3. Surveys would expose, on later analysis, a curious phenomenon. Most Kenyan voters acknowledged that the economy had done better under Kibaki than under Moi. Even a majority of the opposition's natural supporters admitted that their own ethnic community's living conditions had either stayed the same or improved under NARC. Yet there was a widespread perception that particular ethnic groups had done far, far better than the rest, and it was that sense of *relative* deprivation that rankled.[40] 'There's an anti-Kikuyu vote forming, based on perceptions of inequality,' said John. 'Development experts are allowing themselves to be bamboozled by the figures. They are not asking themselves the key question, which is "growth for whom"? The way the rest of the country sees it, it's growth for the Kikuyus, and it hates our guts.' On his travels, John was approached by delegations of diaspora Kenyans, non-Kikuyu who saw him as an ally. 'They tell me they would rather vote in a way that causes an economic slowdown, if necessary, so long as these arrogant Kikuyu are taught a lesson.'

Familiar with the Alcoholics Anonymous programme, John had come to think of his country – so determined to hope for the best, so reluctant to tackle issues that kept bobbing to the surface like a drowned man's corpse – as a boozer in the ultimate state of denial. The first principle of AA is that the hardened drinker cannot cure his addiction until he recognises his problem. Bumbling around, knocking into the furniture, this particular drunk – John was beginning to suspect – would only admit the need for change when he came round in the gutter with puke over his clothes, blood in his hair and his wallet gone. Things might have to get a whole load worse before they got any better.

One of the few to share his fears was KNCHR head Maina Kiai, the other 'anti-Kikuyu Kikuyu'. 'Luos are saying, "No matter what it takes, we must win," and Kikuyu are saying, "No matter what it takes, we can't lose." It's scary.' A few high-profile gestures would be enough to defuse the growing antagonism, but the Kikuyu elite's sense of entitlement was so great it did not see the need. That failure of imagination, Kiai said, had been illustrated a year earlier, when the *Nation* published a photo showing Kibaki shaking hands with all his provincial commissioners. 'All the suits that day were either Kikuyu, Meru or Embu. Every non-Kikuyu saw it, but the people at the *Nation*, who published the photo, didn't even notice.'

But the two doomsayers were in the minority. Most analysts noted how often the country had teetered on the brink of disaster, only for commonsense and the profit motive to triumph every time. Land of the compromise and the fudge, Kenya had a knack for staring ruin in the face, backing off and muddling along. The nation, declared businessmen, journalists and diplomats, had now staged so many multiparty elections it qualified as a seasoned democratic player, with all that implied in terms of stability. Showing the same cheeriness, financial advisers boldly classified the forthcoming polls 'zero risk' –

a remarkable position to take in any African election, let alone one in a nation whose fissures ran so deep.

'The ethnicity thing has actually got much better. The media is just playing it up,' insisted economist David Ndii when we met three months before the polls. I was, he hinted, indulging in a very typical Western stereotyping of Africa. 'Why assume that grievances, even justified ones, translate into violence? We don't take things as seriously as people expect us to. Of course people resent not getting a job because they are from the wrong group, but they know next time around, when it's their tribesman in power, it'll work in their favour. It's part and parcel of our public life. Kenya is a very nepotistic society. We expect it.'

For the first time since John had materialised on the doorstep of my London flat, I found my belief in the importance of what he had done wobbling. My doubts crystallised in, of all places, the slick new Westgate shopping centre, where Nairobi's plaza phenomenon had surely reached its apogee.

I was old enough to remember the tremor of excitement that shuddered through Nairobi when Kenya's first escalator opened in the Yaya Centre in the 1980s, heralding the arrival in East Africa of the modern consumer experience. But Westgate's Nakumatt hypermarket was in an entirely different league. Open seven days a week, extending over two storeys and boasting motorised shopping trolleys for the disabled, its shelves offered everything from dog muzzles to mattresses, artificial flowers to birthday cards, nappies to dishwashers. Forty-five different types of shower gel, twelve varieties of soy sauce, eleven types of toilet paper, this was a supermarket designed to satisfy a fussy society's every whim. Slicing Parma ham and handling French cheese takes certain skills; Nakumatt's staff had been taught them, and they had been drilled to the point where they could recite the precise location of key products without a moment's hesitation. And how Nakumatt's Kenyan customers, wheeling their loaded trolleys to the phalanx of checkouts – each equipped with a plasma screen beaming out slick commercials – revelled in it all, fishing out their loyalty cards and totting up their points.

Wandering the aisles, I suddenly thought that maybe they had been right after all, those determined optimists who insisted it was worth tempering principle for the sake of a greater, long-term good. Oh, I knew Westgate catered for Nairobi's *wabenzi*. I'd seen the nail-varnished Kenyan teenagers impatiently jiggling their Bluetooth mobiles and the keys to the car Daddy had given them at the tills. I understood this gleaming world of dishwashers and catfood bore no relation to life on the *shamba* or in the urban slums. But if this was the way the *crème de la crème* in Kibaki's Kenya shopped, surely the rest of the nation must inevitably be swept along?

I thought of John's warnings. 'Six per cent growth is all well and good, but the trickle-down isn't trickling down.' A small, irreverent voice in my head piped up: 'Well, he *would* say that, wouldn't he?' How could John, who had walked out on this government over a point of principle, bear now to admit that he had got it wrong? How could he accept his sacrifice had been for nothing?

17

It's Not Your Turn

> 'Why worry? It's only an election, not the end of the world.'
>
> Campaign billboard for BROTHER PAUL, aka Kamlesh Pattni

The run-up to Kenya's 2007 elections would see a smattering of small, ominous incidents that should have set alarm bells ringing, if only people were in the mood to listen. One took place in the Rift Valley town of Naivasha, where police seized an assistant minister's white Pajero carrying sixty-nine *pangas* (machetes), two hundred whips and fifty *rungus* (clubs). Another occurred when the Nakumatt supermarket chain announced, after logging a strange sudden spike in machete sales, that it was limiting purchases of gardening tools and kitchen utensils such as knives to just one per person. Some sections of the community were clearly preparing for a fight.

From my perspective, the polls would serve as a litmus test of Kenyan opinion. Did Kenyans care enough about grand corruption to vote out the government, or would they agree with the donors, convinced the 'overall trajectory' mattered more than the inconvenient detail of Anglo Leasing?

That assessment, however, was askew. Top-level sleaze *per se* would not be the issue, as was amply illustrated by Kamlesh Pattni's decision to stand as MP in Nairobi's Westlands constituency. 'Steal a mobile in this country and you get lynched, steal $100 million and you get to run as MP,' scoffed a journalist friend. But the issue of corruption would nonetheless lie at the election's beating heart. As the polls neared, opposition leader Raila Odinga revived a key theme of the constitutional debate, calling for a return to 'majimboism' – a system of devolved regional government left behind by Britain's colonial administration but abandoned by Jomo Kenyatta in 1964. To Western ears, majimboism seemed difficult to fault. Who could deny that power in Kenya, as in most African countries, was damagingly over-centralised? There was learned discussion as to whether the American, German or Canadian federal system would be best suited to Kenya. But for ordinary Kenyans, majimboism meant something very different, and quite specific. To opposition and government supporters alike, the toxic concept challenged the fundamental notion that a Kenyan was free to work, live and invest anywhere in his own country. Under majimboism, those who had bought land, farmed the soil and opened shops outside their ethnic communities' traditional areas would be forced to sell up and move. Given the country's historic patterns of population growth and migration, majimboism's main target could only be the Kikuyu.

This was not so much a principled rejection of the 'Our Turn to Eat' principle of government itself, as a challenge to the notion that one particular tribe should enjoy a monopoly position at the trough. On the stump, Raila's supporters repeatedly cited the Anglo Leasing scandal as concrete evidence of what a bothersome 'certain community' got up to when given half a chance. Ethnic favouritism, the foundation on which Anglo Leasing was built, became the rallying issue of the election campaign. On the websites, non-Kikuyu bloggers sketched a '41 versus 1' scenario: the notion that Kenya's forty-one other tribes must unite, come polling day, against the Kikuyu. I was astonished to hear the same refrain on the lips of a staff member of the Kenyan High Commission in London. 'The way we see it,' he confided in an unguarded moment, 'it's going to be everyone against the Kikuyu.'

'There's a snake in the nest,' Raila declared at his rallies. Using a parable from the animal kingdom, he explained how the safari ant – so tiny, so seemingly insignificant on its own – could, by sheer dint of numbers, overpower the snake that had curled itself around Kenya's hearth. 'You are the safari ants,' he told his audience, to roars of approval.

In the Rift Valley, vernacular Kalenjin radio stations called on listeners to 'clear the weed', to remove the 'spots' on the landscape represented by the 'settlers': Kikuyu whose forefathers had acquired land under Kenyatta. Kalenjin youths, bent on finishing the job started during the ethnic clashes of the 1990s, were more direct. 'Whatever happens in the elections, win or lose, you're out of here,' they told Kikuyu neighbours. The message also won support on the Swahili coast, where local Muslims had for years resented the stranglehold 'upcountry' Kikuyu, who owned many of the large hotels, enjoyed over the tourist industry.

While ODM leaders stoked their supporters' sense of ethnic grievance, the government side – hurriedly mustered under the banner of the Party of National Unity (PNU), yet another ideology-free political formation – systematically whipped up a matching paranoia in Kikuyu ranks. At weekends, hard-line MPs regularly made the trip from Nairobi to Central Province to warn any Kikuyu stupid enough to consider staying home on polling day of the community's looming marginalisation.

When the Kenya National Commission on Human Rights published a report accusing top civil servants of breaching their supposed neutrality and government ministers of misusing public funds in their eagerness to campaign for the PNU, the names cited – from Francis Muthaura, head of the civil service, to Kiraitu Murungi, John Michuki, George Muhoho and Joe Wanjui – read like a roll-call of the Mount Kenya Mafia.

Bit by bit, an ethnic siege mentality was created. 'The notion of defending the government as an institution merged gradually into the notion of defending the president, and that then became defending "one of our own",' recalls *Nation* columnist Kwamchetsi

Makokha. 'It was very, very subtle. If you did a forensic examination you would hardly be able to track the shift.'

As the months passed, Raila, who in 2002 had received a rapturous welcome in Central Province from Kikuyu grateful for the campaigning he had done on Kibaki's behalf, was successfully transformed in their eyes into a terrifying bogeyman. A series of extravagant rumours, inflammatory leaflets and fake memos – Raila was planning to bring in *sharia* law; the ODM had drawn up plans for the genocide of one million Kikuyu; the Kikuyu would be chopped into pieces, Rwanda-style – played their part in creating the mood. 'The amount of fear-mongering SMSs and emails was stupendous,' says Makokha. 'It became a self-fulfilling prophecy. If you set the stage where a single community has isolated itself, what follows is resentment. People start saying, "What's so special about you?" and, "We're going to get these guys."'

Samuel Kivuitu, head of the Electoral Commission of Kenya (ECK), could have been under few illusions as to what was being planned on the government side. In January 2007, violating a ten-year-old agreement that political parties should be consulted on the selection of electoral commissioners, Kibaki unilaterally announced nine new appointments to the twenty-two-member organisation. Then five more. The new arrivals broke with tradition in another worrying way: they chose to supervise the polls in their home provinces, where they hand-picked returning officers. 'I've been trying to train the new commissioners appointed by Kibaki,' Kivuitu privately told EU observers. 'But they tell me they don't need any training, because they are here for only one thing: to take over the ECK.'

There were few more interesting places to be, in the closing days of the campaign, than the western town of Kisumu, in Nyanza province.

Perched on one of Africa's great expanses of fresh water, framed by emerald hills, Kisumu should be a bustling metropolis. The largest city in the Lake Victoria basin, it was once the terminus of the colonial Uganda Railway and should, for no better reason than geographical location, be a magnet for trade from Tanzania, Uganda, Rwanda and Burundi. Instead, it feels like a city time forgot, snoozing in the torpid heat.

The local sugar, rice and cotton industries are either in decline or ticking over, awaiting better days. Kisumu's once-frantic industrial area is now a quiet stretch of shuttered godowns, padlocked compounds and empty yards. The main road to Nairobi is so potholed, only the foolhardy or desperate attempt it. If you value your spine you are advised to go by air, and your plane takes off from outside a glorified shack. A promised international airport has yet to materialise, as has a deep-water harbour.

Even the blue waters of Lake Victoria are not the blessing they once were. Kisumu's fabled mermaid seems to have cast her curse. The bay is choked with water hyacinth, a bobbing green blanket fishermen have to battle across in their flimsy pirogues, which blocks the mouth of the yacht club and hides the herds of hippo the odd tourist comes to see. The fish-processing sector is slowly dying, its workers lured away, managers complain, by Kikuyu factory owners from up-country, promising fat pay packets and fantasy perks.

Kenya's recent economic surge has made little visible impact here. In 2005, according to the government's own statistics, Nyanza overtook North-Eastern as the country's poorest province. Unemployment is rife, and Kisumu cannot even produce enough food to feed its own population. The favourite method of transport – the *boda boda* taxi bike – says a great deal about Kisumu. In other Kenyan cities, cars and vans serve as taxis. But distances here are short, and human sweat is always cheaper than petrol.

Residents harbour few illusions as to the reason for this neglect. 'UK' – 'United Kisumu', as it is known – is the regional capital of the Luo, Kenya's third biggest ethnic group, a community that regards itself as victim of a plot to keep it poor and irrelevant. Two of Kenya's most high-profile assassinations – those of Tom Mboya in 1969 and Robert Ouko in 1990 – were of Luo government ministers. The male flag-bearers of the Odinga family, the family that has been a thorn in the side of every Kenyan president since independence, both spent time in detention. An opposition bastion since 1966, when the then vice president Jaramogi Oginga Odinga, Raila's father, broke with Jomo Kenyatta, Kisumu has paid an all-too-visible price for the family's failure to win their spot at the table. 'These NARC guys never even had a plan for Kisumu or Nyanza province,' says Stephen Otieno, a local community worker. 'Kisumu was never mentioned in any manifesto. You look at that and you start asking yourself: "Is this tribal?"'

But in December 2007, all that looked set to change. Finally, Kenya's game of musical chairs was about to turn in Kisumu's favour. Most opinion polls agreed: Raila was heading straight for State House. Four decades of calculated neglect were about to end, and it would all be thanks to 'Agwambo' ('Man of Mystery'), 'the Hammer', 'the Bulldozer', 'Mr Chairman', a local hero who enjoyed near-god-like status in his own fiefdom. Given Kenya's political tradition of ethnic patronage, a Raila presidency would surely mean new jobs, fresh investment, new roads, hospitals and schools for the Luos, just as it had for the Kikuyu under Kenyatta and the Kalenjin under Moi. On the podium, Raila might insist on presenting himself as a national unity candidate representing all Kenyans. As they queued to cast their ballots on 27 December, Kisumu's residents had a clear sense of what was their due. 'We're voting for change,' was the politically-correct formula. But many quietly added: 'It's our time.'

Turnout, bolstered by the addition to the electoral roll of nearly four million voters, most of them youngsters, was at a record high. For many that meant day-long waits in the sun. But with the exception of the lynching of two policemen suspected of planning to rig the vote – a hint of the suppressed tension in opposition ranks – voting across Kenya was calm. After a day spent touring Kisumu's ballot stations, I returned to my hotel impressed by what appeared to be a quiet demonstration of mass trust in the electoral process. Things seemed to be going smoothly. 'You know, I'm not the sentimental type, but today I really feel proud of my fellow Kenyans,' a journalist friend texted me from Nairobi.

The following day felt equally promising. In defiance of the rule that Africa's elder statesmen, no matter how undeserving, automatically enjoy the respect of the younger generation, the parliamentary results showed that Kenya's irreverent new voters had turned their backs on the dinosaurs. Across the country, government ministers were falling like ninepins: more than twenty would be rejected by their local constituencies. The vanquished included several Anglo Leasing names, Kibaki's pre-election reinstatement proving not quite the canny move he had anticipated. Out went David Mwiraria, the finance minister who had signed off on the dodgy contracts, his concession speech interrupted by calls of 'Loser!', 'Thief!' Out too went vice president Moody Awori – 'Moody will be feeling moody at home,' chortled a radio commentator – and the brawny Chris Murungaru. Out too went Nicholas Biwott, AKA the 'Prince of Darkness', president Moi's sinister right-hand man, and three of Moi's sons, the famous name no protection against the ire of voters disgusted by the two presidents' last-minute marriage of convenience.

With the results from Kenya's western provinces and much of the capital swiftly announced, Raila was well in the lead in the presidential poll and the ODM was wiping the floor with the PNU in the parliamentaries, just as the opinion polls had predicted. But in the Kisumu Hotel's packed restaurant, where diners sat with eyes glued to the television screens, no one dared celebrate. New arrivals slapped hands with friends in a gesture of quiet triumph, but no one allowed themselves more. 'People are hiding their feelings. You can see it in their faces, but they don't want to be proved wrong,' said a friend.

'Any dancing in streets?' a British film-maker friend texted from Nairobi. 'Still too early,' I cautioned. And as the hours ticked by, the atmosphere grew uneasy. Something strange seemed to be happening. The television announcers faithfully relayed one set of figures after another, bludgeoning their viewers with statistics. But where, crucially, were the presidential results from Nairobi and Central Province? Geographically, these constituencies were the closest to ECK headquarters, yet their results were dribbling out at a snail's pace. No one was slapping hands now.

By the morning of Saturday, 29 December, Raila's advantage had been whittled away to the slimmest of leads. Worried guests, gathered around the TV sets, nodded approvingly as ODM leaders in Nairobi angrily demanded to know why ECK was announcing parliamentary and civic results from the Mount Kenya area, while omitting the presidentials. By this stage of the game in 2002, the losing candidate had already been preparing to concede defeat. Why, this time, was everything taking so long?

Kisumu's poor did not wait to hear the answer. To them, it was obvious: once again, the Luos were being royally screwed. The government in Nairobi had conspired to rob their community of its rightful turn at the trough. '*Funga mlango, Funga mlango!*' ('Close the door!') shouted a hotel security guard as we all spotted the same thing through the glass: a ragged mob of youths running towards us, fleeing a police charge. Behind them drifted a pall of dark smoke from barricades of burning tyres. Outside on the streets, the looting had started.

It seemed we were stuck with one another: the Ugandan AIDS worker – a man, by his own admission, 'in the wrong place at the wrong time' – the bewigged mother-of-two, the US-trained entrepreneur with his American twang and techno-savvy son, the angry young law graduate and a handful of tut-tutting staff. Like passengers trapped in a lift, none of us had expected to spend quite so much time in each other's company, but fate had brought us together and refused to let us go.

Just around the corner from Kisumu's main shopping street, the Kisumu Hotel, built in the 1930s to cater for passengers on the Johannesburg–London flying-boat route, turned out to be the perfect vantage spot from which to observe a day-long looting spree. Locked inside by solicitous staff, the hotel's guests – mostly members of the Luo diaspora who had returned to Kisumu to celebrate Christmas and cast their votes – kept up an appalled, sardonic running commentary on the spectacle of a city tearing out its own entrails.

In Kisumu's outlying shanty towns, ODM supporters were taking their fury out on Kikuyu and Kisii residents assumed to have voted for the government, setting fire to kiosks and homes, sending their terrified owners fleeing into police stations and churches. In the city centre, the focus was different. For years, I'd wondered what eventually happened in a society with such a high proportion of jobless, prospectless youngsters. Here was the answer. Some hidden signal seemed to have been passed through the slums – 'Today, anything goes' – and their inhabitants were pouring into the city centre.

First came Bata, Kenya's cheap and cheerful shoe chain. There were so many bulging red-and-white plastic bags on the street, you'd be forgiven for assuming the makers of 'The shoe that says you know Africa' were staging a clearance sale. Then came the electronics stores. Generators, microwaves, ghetto blasters, salon hair-driers. Double-compartmented freezers were herded along the street like cattle, wheels screaming on tarmac. 'And you can bet none of these guys have electricity at home,' mused a guest.

If the looting initially looked random, the impression was misleading. The premises being targeted were all owned by either Asians or Kikuyu. A sitting target, with its glass-paned veranda, our hotel appeared to be protected by an invisible force field. Looters sat on the steps, trying on their stolen clothes for size, but never attempted to enter. 'Why isn't anyone breaking in here?' I asked a clerk, mystified. 'Oh, we're fine. The manager is a Luo. Now, if this was a Kikuyu hotel, they'd already be inside.'

Generous funders of Raila's campaign, Kisumu's Asian businessmen had expected to come through the elections unscathed. They spoke Luo, had lived in Kisumu for generations, saw themselves as loyal supporters of the ODM electoral machine. They had not bothered to take the obvious precaution of running down their stocks. Now they were being administered the sharpest of lessons on the impossibility of integration.

Scores of televisions were wheeled past us on the backs of *boda bodas*. 'Check it out, that one's got a plasma screen . . . isn't that a Super Slim?' asked the US-accented entrepreneur. 'No, not a Super Slim,' his son corrected him. One sensed a heated family debate, back in the States, about which model to buy. The screen promptly crashed to the ground. 'Oh, leave it, just leave it,' muttered the father as the looters struggled to right it. 'Don't you know a dropped TV is always wrecked?'

Next were the furnishings: rolls of linoleum, fake Persian carpets, sheets of corrugated iron, entire velour sofas, balanced on heads or carried between two men, piled high with goodies. 'Terrible, just terrible,' muttered the matron, lifting her daughter up to give her a better view.

The local MPs whose victories had recently been announced were nowhere to be seen – they had no intention of putting their fragile authority to the test. And events in distant Nairobi, relayed over the television and radio, were doing nothing to calm this frenzy of impotent rage. Rumours circulated that the chief justice had already been called to State House, on standby for an imminent Kibaki swearing-in. At a tumultuous press conference Kivuitu admitted that the delay in announcing results was 'unacceptable' and revealed that he had lost telephone contact with many Central Province returning officers. What, speculated viewers, could these key officials possibly be doing, apart from conjuring up the extra votes needed to ensure a Kibaki win? 'We don't know where our returning officers are,' confessed Kivuitu. 'In State House!' called a member of the audience.

Watching the chaotic scenes on TV, the law student from Mombasa waved at the screen in contempt. 'This is black democracy,' he pronounced with terrible bitterness. 'It's giving me an ulcer,' he muttered, departing for his room.

Occasionally an army truck drove by, firing teargas in a desultory fashion, and everyone scattered. But it was like trying to dam the tide: five minutes later the looters surged back. A curfew was desperately needed, but none was announced. Arriving guests said the security forces had been spotted shooting the locks off Asian properties and stopping looters in order to claim a share of the booty. This looked more like collusion than incompetence. What, after all, did it matter to the powers that be in Nairobi if the Luos fouled their own nest?

We began recognising faces on the street. The same looters were coming back for second, third and fourth helpings. Some were children. A surprising number were women. Two strapping Amazons, running along in kitten heels, breasts bouncing, wigs askew, made a strong impression. 'These are thugs, women thugs,' said the hotel clerk with a shake of his head.

On it went, hour after hour. Sure of their ground, tired looters were now strolling along the main drag, not bothering to run. Much of the centre of town was aflame, sending angry charcoal clouds billowing into the air, gas canisters exploding like bombs in the heat.

As the sun set, the looters finally ran out of energy. Their occasional 'No Raila, No Peace' chants had come a definite second to the most vandalistic operation of self-enrichment I'd ever witnessed. In the space of twelve hours, Kisumu's shopping district had come to resemble a war zone, its shops reduced to blackened husks, its streets crunchy with glass. Carrying out the task their government in Nairobi was so unwilling to perform, the *wananchi* had done their bit to redistribute Kenya's assets. The laborious 'trickle-down' so beloved of World Bank economists had been replaced by the far quicker smash-and-grab. 'It's not your turn,' the Kibaki elite had told them, and they had shown exactly what they thought of the message.

So much for the trouble-free election. 'In future, I will never, ever write about a country being "stable" when so much of its population lives below the poverty line,' commented a Dutch colleague in Nairobi to whom I was relaying events. Behind me, a hotel security guard mused to himself as he watched the stragglers trudging back to the slums, 'When you go home and you have looted like this, what do you think to yourself?'

Long after they could have played any useful role, riot police in helmets deployed across the smouldering centre. The hard work would come the following day, when they would ruthlessly exert their authority over Kisumu's slums, shooting without discrimination. I heard one of my fellow guests, in surreally chirpy mode, on his mobile phone to a friend. 'Hey, don't bother coming shopping down here tomorrow. There's nothing left to buy.'

* * *

The rape of Kisumu city centre was just a taste of what was to come.

By Sunday, 30 December, Kenya's elections were a fiasco, the country's worst crisis since independence. There was growing evidence of ballot-stuffing, glaring disparities between the figures announced by returning officers at constituency level and what the ECK was releasing in Nairobi.[41] There had almost certainly been ballot-stuffing in the ethnic strongholds of both the ODM and the PNU, but since the outcome looked set to be a Kibaki win, all eyes were focused on the government side.

At a last press conference in Nairobi, furious opposition leaders stormed the stage, determined to prevent the harried Kivuitu announcing the results. The GSU brought an end to the event by turning off the power and firing teargas. The final tally, which the chairman read out to a camera from the side room where he had taken refuge, demolished Raila's seemingly unbeatable advantage. In the closest electoral race in Kenya's history, Kibaki had won by 231,728 votes. The result not only flew in the face of pre-electoral opinion polls, it jarred with parliamentary results giving the ODM ninety-nine seats and the PNU just forty-three.

'If it was up to me, I wouldn't sign off on this,' Kivuitu had privately told EU observers, showing them returning officers' forms bearing obvious signs of tampering. 'But I'm alone here, and I want to live.'

Less than an hour later, Kivuitu reappeared on Kenyan television screens, this time witnessing an inauguration on the State House lawn so swiftly arranged conspiracy theorists immediately speculated that it had actually been filmed ahead of time. Staged in the absence of the Kenyan public, neighbouring African presidents and the diplomatic community, the surreptitious ceremony was in the starkest possible contrast to Kibaki's gloriously chaotic 2002 inauguration in Uhuru Park, when the nation had come together. 'Like a Muslim funeral, the swearing-in of Mwai Kibaki took place before even some of the relatives of the deceased knew what was going on,' was the sardonic comment of Ugandan columnist Joachim Buwembo. The head of the Institute of Education in Democracy, Koki Muli, captured the feelings of many Kenyans. 'This is the saddest day in the history of democracy in this country. It is a *coup d'état*.'

In the days that followed, independent election-monitoring teams from the EU, the East Africa Community, the Commonwealth and Kenya itself lined up to denounce the vote-tallying process and slam the ECK, whose reputation was not helped by Kivuitu admitting he 'did not know' whether Kibaki was the real winner. But by then the election's flawed conduct would almost seem a mere detail of history, such was the magnitude of what was happening on the ground. Within minutes of the announcement of Kibaki's victory, the multi-ethnic settlements of Nairobi, Mombasa, Kisumu, Eldoret and Kakamega erupted. Luo and Luhya ODM supporters armed with metal bars, machetes and clubs vented their frustration and fury on local Kikuyu and members of the smaller, pro-PNU Meru, Embu and Kisii tribes, setting fire to homes and shops.

The approach was brutally simplistic. Many Kikuyu, especially the young, urban poor, had actually voted ODM, regarding Raila, 'the People's President', as far more sympathetic to their needs than the aloof Kibaki. But mobs don't do nuance. Fury needs a precise shape and target if it is to find expression, and ethnicity provided that fulcrum. The attackers claimed they took action because they could not bear the sound of Kikuyu celebrating 'their man's' victory. But their victims said they would have never been so brazen or so foolish, and pleaded with fellow slum-dwellers to remember they were all poor people, suffering together.

Some Western reporters wrote of 'atavistic tribal tensions' bubbling to the surface, implying the hostility between Kenya's communities was a mindless, irrational thing. But under a system which decreed that all advancement was determined by tribe, such hostility was entirely rational. Had all Kenyans believed they enjoyed equal access to state resources, there would have been no explosion. As Bill Clinton said in another context: 'It's the economy, stupid.'

This violence was horribly up-close and personal. In Korogocho, Mathare, Dandora and Kibera, neighbour raped neighbour, husband murdered wife, schoolmate killed schoolmate. As Wangui Wa Goro, a London-based commentator, put it: 'When you're unleashing decades of frustration, you're not going to hunt out some stranger down the road. You're going to go for the noisy woman upstairs who's been driving you crazy for years.'

There were terrible echoes of the Rwandan genocide in the tactics involved. ID cards in Kenya don't show tribe, but they give paternal birthplace and family name, often ethnic giveaways in themselves. At makeshift roadblocks set up on Kenya's main thoroughfares, passengers were forced to descend from *matatus* and buses and to show their ID cards. They were then beaten or killed if they belonged to ethnic groups deemed likely to vote PNU.

In the Rift Valley, Kalenjin warriors systematically torched Kikuyu *shambas* on the hillsides, the farmers fleeing before the chain of fire reached their compounds. Issued with arrows, spears and jerrycans of fuel by local elders and community leaders, Kalenjin youths were ferried around in trucks, evicting hundreds of thousands of so-called 'foreigners' – many of whom had never lived anywhere else. When their victims hid in a church they set that ablaze too, burning women, old people and babies inside.

Brushing aside the official election result, ODM's supporters had collectively moved to put their majimboist interpretation into practice on the ground. Challenging the very notion of the unitary state, they set about systematically reversing decades of Kikuyu expansionism. It was no coincidence, commented British historian David Anderson, that 95 per cent of the Rift Valley clashes occurred in areas where land had been distributed as part of Kenyatta and Moi's notoriously corrupt government settlement schemes.[42]

Kenyans had expected a rocky election, but nothing on this scale. Before the poll, I'd spent an afternoon with a garrulous young painter in Mathare slum. Bubbling with political convictions, he'd treated me to a warm Fanta in his 'seven-in-one', the windowless shack that served as sitting room, artist's studio, study, bedroom, storage facility and kitchen. A Kikuyu in a largely Luo section, he sold his works to the international market, but was proud of never having left Mathare. He had stowed his most valuable possessions with middle-class friends, but was confident he could see through any coming storm. He ran painting workshops for slum children, played a high-profile role working for community integration and had more Luo friends than Kikuyu. 'If trouble comes, I'll just put on my ODM shirt,' he said with a relaxed shrug.

I met him a week after the elections, camping with his siblings in a friend's suburban villa. They had all been forced to flee Mathare. His ebullience had evaporated. He would never, he said, make the mistake of voting across ethnic lines again, for none of his Luo acquaintances had returned the favour. 'I'm becoming more tribalistic with each and every day. In future, wherever I go to live, I will want to know who my neighbour is on my left and who is on my right, and if we are ten and there are only five Kikuyu, I will want to bring in another Kikuyu. That's the way we will think in future in Kenya.'

He rang me after I took my leave, his voice flat. Before, he had been happy – proud – to be quoted by name. Now he said, 'Please don't use my name. Someone who spoke to the press has just been hacked to death in Mathare.'

The election results had acted like a sharp tap on the side of a long-damaged porcelain vase, suddenly revealing an already-existing network of hairline fractures. A painful shift in perspectives had occurred, a fundamental recalibration that caught most Kenyans floundering, looking strangely naïve. Before, tribe had been something to whisper over, joke at and bitch about. Suddenly it was the only thing that mattered, and everyone found themselves wrong-footed, stumbling over the new terrain.

Employers scrambled to charter planes to perform short-haul loops around the country, scooping up employees whose ethnicity, irrelevant when they were appointed, now meant they were in mortal danger. 'This sounds ridiculous, but we've somehow managed to send a Kikuyu camera crew to Kisumu,' one television producer told me. 'Of course they can't leave the hotel, let alone do any filming, so they'll have to be replaced. I'm used to taking these issues into consideration in other African countries. But not here. Not in Kenya.'

Nowhere was this dawning of ethnic self-awareness more sudden than in the slums, Kenya's melting pots, where new frontiers coagulated like DNA strands, forming as suddenly on the ground as they had in people's minds. The notion that urban youth would serve as midwives to the birth of a cosmopolitan, united nation looked like idealistic nonsense – the worst violence took place in places like Kibera and Mathare, and it was committed by youngsters.

For crime reporter Robert Ochola, the experience had the vividness of a lightning strike. 'I was in Huruma, where there's a huge barrier – not a physical one, an imaginary one, an equator,' he recalled. 'I just happened to be on the Kikuyu side, collecting reports, when someone came up and said, "Hey, you're a Luo." It hadn't occurred to me until that moment. My heart started racing, a crowd began gathering, I could see the flash of pangas here and there. If it hadn't been for those I was with, I'd be dead by now. I have to accept, now, that I'm a Luo first and a journalist second.'

But similar double-takes occurred in Nairobi's gated communities, where the middle classes, who had assumed they had brushed off such distinctions along with the mud of the upcountry *shamba*, discovered a new social awkwardness. Forget Sheng, there was now one conversation for one's own kind, another for outsiders. 'I went to dinner with colleagues recently and there was silence round the table,' one young woman told me. 'We were so aware of the landmines in the conversation – because it was a mix of ethnic groups – no one dared say anything.'

We were all shamefully aware of sprouting new, ethnic antennae. In my address book, I found myself surreptitiously scribbling 'Luo', 'Kik' or 'Kam' after entries. As the violence escalated, my contacts were being forced to choose their camp. Formerly nuanced, their positions were growing cruder, their opinions shriller by the day. To interpret what someone said, you needed to know who they were.

'Save our beloved country,' Kenyan newspapers pleaded, calling on Kibaki and Raila to talk. But the media's message itself was muffled and blurred: weeks into a crisis taking the crudest of ethnic forms, journalists stuck stubbornly to their self-imposed rule of never identifying tribe. Kenyans might be killing each other, slum-dwellers occupying one another's shacks and entire neighbourhoods upping sticks on purely tribal lines, but the local media still coyly refused to tell their audience who was doing what to whom.

No one, in any case, was listening to the press. Having refused to surrender State House, Kibaki was bent on entrenching his position. Much of the eventual 1,500 death toll could be laid at the door of the government, which announced an unnecessary ban on public demonstrations and then ordered the mainly Kikuyu GSU, issued with live ammunition, to ruthlessly enforce it.

Taking the wind out of Africa Union mediation efforts, Kibaki named a partial cabinet whose members, among them Kiraitu Murungi, George Saitoti, John Michuki, Amos Kimunya and Martha Karua, included not only a host of Mount Kenya Mafia hardliners, but any survivors of the Anglo Leasing and Goldenberg scandals who had managed to hold on to their parliamentary seats. Having tumbled into the ditch, the drunk had hauled himself to his feet and headed straight back to the bar to order another beer.

Five years of economic recovery were sabotaged in a matter of weeks. Tourists packed up and fled, whisked away by charter companies which then cancelled future flights. With militias blocking the roads and ripping up sections of the colonial railway, goods could no longer cross the country. Flowers lay rotting in silent airport storage rooms, maize stood unharvested, vegetables failed to get to market and Kenya's landlocked neighbours watched aghast as fuel deliveries ground to a halt, crippling their own economies. Seeing their hard-earned profits shrivel, appalled Kikuyu professionals formed lobby groups, drew up detailed peace plans and called on Kibaki to extend a conciliatory hand. 'Kibaki is entirely to blame for this,' said one banker friend. 'Raila was not a difficult person to satisfy. By his absolute indolence, Kibaki failed to manage a pretty easy process. And now, because of what he did, we're hated. Oh boy, are we hated.'

The response was silence. It was others' turn now to share John Githongo's revelation. 'People are beginning to realise it's not a question of Kibaki being misled by the hardliners around him,' one investment expert told me. 'He *is* the hardliner.'

Fury at the danger to which the Mount Kenya Mafia's greed had exposed their community was not shared by all Kikuyu. Instead of lambasting their leaders for playing the ethnic card, many actually thanked them for saving them from the ODM bogeyman, who had now assumed such proportions his monstrous outline seemed to blot out the sun. 'I keep getting emails from friends saying, "We all know Kibaki rigged." Everyone knows,' said a musician friend. 'And you know what? We're all so glad!'

At the end of January 2008 came the Kikuyu backlash. For weeks, rumours had circulated that PNU hardliners were raising funds and mobilising the Mungiki, a Kikuyu criminal organisation usually confined to the slums, where it ran protection rackets and oversaw the illegal brewing business. This army for hire now went into action in the lake towns of Nakuru and Naivasha. When British journalist Lucy Hannan found herself driving past hundreds of young Kikuyu fighters assembling on the outskirts of Nakuru, she noticed a small, telling detail. 'I realised after we had passed them they were all carrying brand-new machetes. That's where the Nakumatt machetes ended up.' This was ethnic solidarity at its ugliest. In full view of the police, the Mungiki crudely circumcised non-Kikuyus with broken bottles before beheading them. When groups of Luos and Kalenjins retaliated, it took army helicopters, firing from the sky, to separate the two sides.

As ever in Kenya, those with the least paid the highest price. On one side of the police cordons, Kenya's middle classes were paranoid but protected; on the other, slum-dwellers slaughtered one another. 'Oh, don't worry,' I heard one resident sardonically telling a friend calling from abroad for reassurance, 'here in Nairobi's gated communities we have staff to do our looting and raping for us.'

By mid-February, with some 300,000 displaced Kenyans living in camps and another 300,000 on the move, analysts were speculating about how long it would be before the army took over. The worst bloodletting in the country's brief history had destroyed its always misleading image as one of the few 'tame', 'user-friendly' African destinations. The nation was in shock, though many recognised their surprise was impossible to justify. 'It's bizarre,' pondered one foreign correspondent. 'We've all been writing for years about ethnic tension, the growing divide between rich and poor, unsustainable pressure on the land. And yet somehow none of us digested the implications of our own articles.' After years of being made to feel inferior, Kenya's neighbouring countries luxuriated in the feeling of *schadenfreude*, talking sagely about 'learning the lessons' and 'not going down the Kenya route'. Interviewed by a Nairobi newspaper, Daniel arap Moi could not resist the chance to crow. 'I told you it would be disastrous . . . I okayed it because you insisted on democracy. But let me ask you, is arson the new democracy you were talking about?'

In the space of only two months, Kenya had changed beyond recognition. Rolling back the migration trends of half a century, a process of self-segregation was under way. 'You have a right to reside anywhere in Kenya,' shouted the red headline on a government statement published in the newspapers. But no one believed that now. Kikuyus, Merus and Embus flooded back towards Mount Kenya, Luos, Luhyas and Kisii streamed westwards, Kambas headed east. Teachers abandoned their schools and moved to areas where they felt safe, only to find many of their pupils had beaten them to it. University students and their professors applied for course swaps and transfers. Landlords gave tenants notice on the basis of ethnicity – 'Oh, we just can't trust them any more' – flower farmers fired pickers to open up jobs for kinsmen, and kiosk-owners asked customers for ID before handing over groceries. Even the health system showed signs of Balkanisation, with ODM supporters checking into hospitals where they were sure not to be treated by Kikuyu staff. The voluntary zoning, first symptom of national disintegration, took place to begin with irrespective of class and income.

Kenya's crisis could be encapsulated in a single archetypal image: a Toyota pick-up, piled high with mattresses, a chest of drawers in one corner, bed frame in the other and a medley of pots, pans and plastic bowls in between. Thousands of families were criss-crossing the country, returning to what were optimistically being dubbed their 'ancestral homelands' but were, in fact, areas where most had no land, no assets, no friends and precious little family.

Former UN secretary general Kofi Annan, the emissary of a panicking international community, flew in to mediate between government and opposition. But it was not until 28 February, as Kenya stood braced for a Round Two that everyone knew would make Round One look like a gentle skirmish, that Kibaki finally blinked. Reviving a post phased out in 1964, he agreed to share power with an executive prime minister, concede key cabinet positions to the opposition and investigate the flawed election. The irony was that, combined with undertakings to review the constitution and discuss sweeping electoral, parliamentary and judicial reform, the deal brokered by Annan and Tanzanian president Jakaya Kikwete contained most of the elements of NARC's pre-2002 programme. Had the regime only delivered on its original promises, Kenya could have been spared a multitude of horrors.

Was Kibaki swayed by the realisation that his nation was heading towards civil war and economic ruin? Sadly, no. He folded for entirely self-serving reasons, in the face of a chorus of increasingly explicit threats. Warning that it would not be 'business as usual' if a political deal went unsigned, Western embassies leaked plans for a series of travel bans to be slapped on regime hardliners. US secretary of state Condoleezza Rice flew in to deliver the message that her government was ready to find and freeze the assets of Kenya's leadership, just as it had with pariah nations like Iran and Sudan. The final straw was a simple message from Kenya's generals: 'If you don't clean up this mess, the army will.'

Traumatised Kenyans were then subjected to the distasteful spectacle of their elected representatives wrangling over cabinet posts and attendant perks. 'We will not eat bones while the others are eating all the steak,' said one ODM official, illustrating that Kenya's near-death experience had done nothing to alter the elite's traditional view of politics. With forty-two departments, the coalition government finally unveiled was the largest in Kenyan history, accounting for 40 per cent of MPs. Merely servicing this bloated exercise in jobs-for-the-boys would gobble up 80 per cent of the national budget, Mwalimu Mati's anti-corruption group estimated, forcing cuts in spending on health, roads and education. When reformers had called for a system which ensured all got to 'eat', this was not exactly what they had intended.

The mantra now in Kenya is that the nation peeked over the abyss, saw the scattered bones and burning shacks below, and drew back just in time. 'We have been to hell and back,' prime minister Raila Odinga told British parliamentarians, '[but] Kenyans now finally know and understand one another.' This is almost certainly wishful thinking. Kofi Annan's marriage of convenience is likely to last only as long as it takes the political players to build up war chests ahead of another electoral bout. Some 150,000 Kenyans still languish in squalid camps for the displaced, gradually turning into slums as tarpaulin is replaced with durable iron sheeting. Too frightened to return to contested *shambas*, with no way of earning a living, they are easy recruits for the fundraisers who believe the militias will deliver the solution to the questions the politicians skirt. Being initiated into violence is like losing one's virginity: innocence cannot simply be wished back into existence. These youngsters now understand the alternative use of the *panga*. Momentarily dampened, their hatred only awaits the next referendum or election to re-emerge, red-eyed and bent on revenge.

Annan's peace deal catered for a series of commissions to probe not just the election's conduct but the deeper causes of the violence, including Kenya's festering land disputes, its human rights abuses and its off-kilter constitution. The work is long overdue and desperately needed, but Kenyans know their country has a history of commissions whose findings and recommendations are either watered down before released or simply ignored.*

* At the time of going to press, the Kriegler and Waki commissions had delivered their reports, the first recommending the radical reform of the Electoral Commission of

While businessmen talk bravely of rebuilding shattered town centres and ministers call on tourists to return, many suspect the damage done the national psyche cannot be repaired. It will take a generation, at least, for young Kenyans to forget the images of slaughter. 'The generation that harboured that kind of ethnic hatred was dying away,' says John Kiriamiti. A former bank robber, he renounced crime to become a respectable newspaper publisher in Muranga, and now quails at the violence he once took in his stride. 'Our children didn't know about it. But they have understood it now, and it will take a long, long time to vanish.' The myth of Kenyan exceptionalism – the notion that the chaos associated with other parts of Africa simply 'didn't happen here' – has been forever laid to rest. Kenya has become a land where bruised ethnic communities, whether Luo, Kalenjin or Kikuyu, wallow in the conviction they have been supremely hard-done-by, while striking terror into the hearts of equally aggrieved ethnic rivals. 'They have opened the Pandora's box and let all these issues out,' an Asian shopkeeper in Nairobi told me with a shake of his head. 'It's hard to know how they can ever close it again.'

John Githongo has been proved correct in the most terrible way. Long before most of his Kenyan contemporaries, he recognised graft's awesome potential to destabilise and destroy a society. There could have been few more lurid illustrations of the fact that government corruption, far from being a detail of history, really does matter, than Kenya's post-election crisis.

Kenya and the second calling for the senior figures who instigated the election violence to be either tried before a special tribunal in Kenya or prosecuted by the International Criminal Court in The Hague. A new electoral commission looks certain, but the Waki report, which has huge political ramifications, is yet to be acted upon.

EPILOGUE

'This is the history of a failure.'

CHE GUEVARA, *The African Dream*

On 16 July 2006, a young man died in a district hospital in Narok, a dusty town on the road to the Maasai Mara, Kenya's most frequently visited safari park. He was only thirty-eight, and the illness which killed him is treatable. But poor Africans, receiving only spasmodic medical care, often die of ailments that would be beaten off in Western Europe. He left behind a widow and three children.

The man's name was David Munyakei, and in a curious way he represented John Githongo's *alter ego*. Of a similar age, the two men met and shook hands only once, at a prize-giving ceremony held a few months before John went on the run. One wonders whether either, at that moment, sensed their strange kinship.

Like John, Munyakei was a whistleblower, a man whose rigid sense of right and wrong made it impossible, at a key point in Kenyan history, for him to remain silent, even if speaking out meant losing his job – and worse. Like John, his actions exposed a multi-million-dollar scam reaching to the highest echelons of government. But there the similarities end. Munyakei, unlike John, did not belong to Kenya's upper class. An illegitimate child, he was born in Langata women's prison, where his mother was employed by the Prisons Department. In contrast to John, a highly educated professional confident of finding well-paid employment wherever he chose, Munyakei spent his life worrying about money. While John was able to flee abroad, Munyakei's horizons were necessarily smaller: he simply disappeared inside his own country. John, a natural charmer, could call in favours across the continents when he reached his point of no return. Munyakei, a diffident, not particularly likeable young man, found himself terribly alone, fighting forces poised to crush him. This was a man who had no safety net, no fallback position. And his eventual fate was far more typical of most African whistleblowers than John's.

Munyakei had considered enlisting in the army, but in 1991, when he was twenty-three, he joined Kenya's Central Bank instead. A bit of string-pulling brought the offer of a clerk's job, and he moved to the department responsible for pre-shipment export compensation. As part of a government scheme to increase foreign exchange reserves, exporting companies at that time benefited from a generous compensation scheme. The CD3 customs forms companies submitted to secure payment, declaring goods for export, went through Munyakei's hands, and he began noticing irregularities. 'I could see that I was processing the same forms again and again. The numbers had been changed – they'd been whited over and filled in again – but everything else was the same. And the sums being paid out were enormous, enormous.' Billion-shilling payments were common, and the forms were arriving two to three times a week. The other thing that made Munyakei suspicious was the way his working routine changed. 'These transactions used to take place at very odd hours – after five p.m., when no one else was around. The head of my division asked me to stay behind. He would present me with the documents and say, "I want you to work on this particularly." I was not to go home until the respective accounts were credited.'

Munyakei was a tiny but vital cog in a machine that was eating up billions of shillings in taxpayers' money, supposed compensation for non-existent gold shipments. Processing the payments, he became increasingly uneasy. One Central Bank official, who had himself refused to sign off on the payments, warned him: 'These things are not proper. You must take care.' Colleagues from other departments ribbed him – 'Mister, where exactly is the gold you seem to be clearing?' But his superiors fobbed off his enquiries. Instructions came from above, they said; these were not the concern of a lowly clerk.

What his bosses hadn't catered for, just as John's colleagues never spotted it, was the element of sheer mulishness in Munyakei's character. He might not have attended some expensive Western university, but he was nobody's fool, and he didn't appreciate being treated like a patsy.

One got a glimpse of that streak of truculence within seconds of meeting him. Maybe the long years of being snubbed were to blame, but there was no conversational give-and-take with David Munyakei. Ignoring all interjections, he talked fixedly at you, finger jabbing the air, eyes locked unblinkingly onto yours, frowning in concentration. One sensed that he might not have been an unmitigated joy to work with. Slight, with badly pocked skin, he came across as prickly and awkward, the kind of civil servant who would be a stickler for detail, whose professional inflexibility might drive more easygoing workmates to distraction. This, most definitely, was not a man who would simply shrug his shoulders and mutter the words used to justify wrongdoing from time immemorial: 'I was only following orders.'

The scandal exploded into the public arena in April 1992, when the *Daily Nation* published a series of articles about Goldenberg, the company benefiting from this generous compensation, and shady dealings at the Central Bank. A year later, Munyakei was arrested at work and taken to CID headquarters. He was charged with violating the Official Secrets Act. Munyakei was not the only employee unhappy at what was going on, but his superiors had decided he was responsible for leaking Kenya's largest ever officially-sanctioned financial scam, photocopying sensitive documents and spiriting them out of the Central Bank. On hearing the news, Munyakei's shocked mother suffered a stroke, dying without regaining consciousness.

'I did it as a patriotic citizen,' Munyakei protested. 'I knew what I was doing' – a touch of pride there. 'I was not guessing, I had the evidence. I thought I was doing the government a favour. Instead we got arrests, court cases, threats and dismissals.'

Five months later, the attorney general ruled that Munyakei had no case to answer. But his ordeal had only just begun. The Central Bank sacked him, and refused to give him references, making future employment in the banking sector impossible. As the government sought to placate outraged donors by going through the motions of investigating Goldenberg, Munyakei started receiving threatening anonymous phone calls. 'I would meet people from the bank, and they would say to my face: "It's better if you disappear from Nairobi completely. If you keep pressing the bank to reinstate you, you may not live long."' He fled with a cache of sensitive documents to Mombasa, where he reinvented himself, taking a new name, converting to Islam and marrying a local girl. Eventually the jobs ran out, and Munyakei moved his new family upcountry, to the shack his mother had built on a windswept homestead in Maasai country. The white-collar worker became a gum-booted subsistence farmer, eking a pitiful living from the dry land, just as his ancestors had done. The Goldenberg scandal simmered inconclusively on, with donors nagging president Moi for action. Kenya had forgotten about David Munyakei.

Then came the 2002 election, and NARC's establishment of the Goldenberg Commission. It felt as though fate had reached out and tapped Munyakei on the shoulder. Testifying before the Commission's judges, lawyers and financial experts, he finally won his moment in the sun. A lawyer remarked that if ten witnesses could only be of an equal standard, the Commission's work would soon be done. There were hotel receptions and photographs, the Transparency International award ceremony at which he met John Githongo. The reverberations rippled all the way back to Narok, where Maasai elders materialised on his doorstep asking for a share of the booty. He had to show them the door. None of this notoriety brought in any money. Back home he was still just a peasant, hoeing his plot and struggling to pay his girls' school fees. The one thing Munyakei really wanted – his Central Bank job back, with compensation for years of lost earnings – remained out of reach. He had betrayed his colleagues, he was told, and they would never work with him again.

With the release of the Bosire report into Goldenberg, his hopes briefly lifted, only to be dashed. 'Did you see how few mentions there are of me?' he said, jabbing the report. 'Did you see? I'm very disappointed.' The report noted that there was no legislation protecting whistleblowers in Kenya, but did not suggest how this lacuna could be addressed. There was no reference to compensation. 'The people who face prosecution are very rich and powerful. They could do anything to ensure people don't give evidence in court. We've seen this before in Kenya. Not one of the witnesses in the Ouko murder case is living now. They all died mysteriously. I feel very insecure, now, very much demoralised.'

But my impression was that what really dismayed Munyakei was not his personal security, but the fear of going down in history as a fool. He craved a ruling that would validate his act in his own eyes and those of his community. 'I thought as a whistleblower you should be treated with high esteem. But it appears,' he scowled, 'that when you do such a thing no one appreciates it. If nothing is done for me, no one will come forward and expose corruption in future. They will look at me and say, "Munyakei did it, and what did he get? He ruined his life."'

He got up to leave. Could I give him something, he asked, for the taxi fare at least, or the cost of his Nairobi hotel? Times were tough in Narok, he explained. The drought was killing crops and cattle. The man belatedly hailed by government ministers as a national hero was struggling to make ends meet.

I had planned to visit him in Narok, but left it too late. His nervousness proved misdirected. Future Goldenberg prosecutions were robbed of a key witness not by a hired assassin but by an ordinary African killer: pneumonia.

The tale of David Munyakei reads, in many ways, like a practice run for the John Githongo story. John was able to improve on Munyakei's dress rehearsal in part because of his status and birth. Munyakei could not muster journalists in Kenya and Britain to publicise his cause, could not tap a network of academics, foreign institutions and VIPs for succour and sympathy. But John had also learnt explicit, valuable lessons from his predecessor, in particular when it came to collecting the evidence that proved he was no fantasist. 'Can you imagine how I would look now if I hadn't taped those conversations?' says John. 'Just some loony in Oxford making crazy claims.'

In that progression lies a clue to Africa's future.

Both men set out to end Kenya's long tradition of impunity by making a ruling regime accountable for a major corruption scandal. Neither succeeded. To date, only the lowly have faced legal pursuit. In Anglo Leasing's case, three permanent secretaries and three junior civil servants prosecuted in 2005, on John's watch, have yet to be tried. Justice Ringera told me that until investigators Kenya asked for help in France, Switzerland, Britain, the United States and the Netherlands have provided details of the foreign firms and bank accounts involved – and, sadly, that might take some time – he could not send his files to the attorney general for prosecution. With Goldenberg, the dropping of charges against high-profile figures continues. Since the suspected perpetrators of both scams hold key positions on either side of the political divide, each faction can effectively blackmail its rival. The culprits are likely to spend the rest of their lives being driven around by chauffeurs, relaxing in Nairobi's wood-panelled clubs and strolling the lush fairways of its golf clubs.

But prosecutions are not, actually, the point. What is remarkable is that first David Munyakei and then John Githongo chose to launch their respective Missions Impossible at all, taking on their own society's 'Our Turn to Eat' culture and defying the rule which dictated that loyal employees might know exactly what tricks their bosses were up to, but could always be relied upon to remain silent. The contents of John's dossier and Munyakei's testimony matter far less than the fact that they emerged in the first place to challenge the system. As the playwright and Czech president Vaclav Havel said: 'Even a purely moral act that has no hope of any immediate and visible political effect can gradually and indirectly, over time, gain in political significance.'

Mention John Githongo and his dossier to an educated Kenyan and you will often get a raised eyebrow and a sardonic look which suggests that Westerners would do well to abandon their naïve fixation on lone heroes. 'What he did is of no significance. None,' a yuppie businessman proud of his opposition sympathies surprised me by saying. Even admirers will tell you that the number of Kenyans who think like John is so small and atypical their words and actions can have no fundamental impact on larger society. Their scepticism is shared by a school of Western analysts who see sleaze in Africa, intertwined as it is with cultural respect for the extended family and ethnic loyalty, as part of the continent's very haemoglobin. In these commentators' view, the grand corruptors will always be able to remodel their looting techniques to keep one step ahead of those trying to purge the system, in part because they enjoy as much public admiration as they do opprobrium.

They are wrong to be so dismissive, or so despairing. Cultural values are not immutable; they shift all the time, as the West's own history – witness, for example, the view our courts take today, as opposed to fifty years ago, of a woman's rights in marriage, or of racism in the workplace – amply demonstrates.[43] The dramatic changes Africa has experienced in the last hundred years, hurling its citizens from herding livestock on the *shamba* to lunchtime at the cyber café, shows the continent is no bizarre exception, impervious to the trends and processes that affect the rest of humanity.

John Githongo is no saint, as his many exasperated friends can attest. But the fight against graft in Africa is not to be settled in one battle, or by one person. It is being waged in stages, step by step, with many a sideways shuffle and backwards totter, by many individuals. By doing what he did, John, like David Munyakei before him, permanently shifted the debate's parameters. Altering expectations of how a civil servant under pressure could behave, he made it possible not only for others to follow in his wake but to move beyond him. Everyone needs role models, helping them see the way ahead. 'I'd like to throw a small spanner in the works,' John says. 'I'll do my little bit and the next time it'll be someone else and someone else and someone else. At the very least, it should never again be possible for civil servants and politicians to get together in a room and discuss how to rip off the Kenyan people.'

Corruption relies for its survival upon an intangible, unspoken agreement of what is tolerable behaviour – witness the checkpoint policeman's reluctance to ask *mzungu* drivers for bribes 'because they will make a fuss', which prompts the obvious question: 'Well, why don't the *wananchi* make more of a fuss?' And while they grabbed the headlines, David Munyakei and John Githongo are not the only ones challenging such tacit acquiescence. Maina Kiai may have left the KNCHR – 'They have killed this country, killed it. I won't be applying for another term,' he told me during the election violence – but anti-graft campaigners Gladwell Otieno, Lisa Karanja and Mwalimu Mati, whose remarkable Mars Group website features a grim 'inactivity counter' logging government obfuscation, still fight on. Writers and bloggers like Parselelo Kantai, Muthoni Wanyeki, Binyavanga Wainana, Ory Okolloh and Martin Kimani, and a host of civil society activists, human rights campaigners and clergymen are changing the nature of what is considered acceptable in modern African society. The time taken for grand corruption to be exposed to public view, if not punished, has concertinaed: 'Goldenberg took forever, Anglo Leasing has been far shorter, and the next one will be even quicker,' predicts constitutional lawyer Wachira Maina. And Munyakei's story shows that you do not have to belong to the elite to rock a system by just saying 'no', although it certainly helps.

What role should Africa's foreign partners play in this battle? This book does not seek to argue that donors should cut all aid to Africa, on the grounds that 'It'll only be stolen,' as the cynics claim. It does, however, hope to alert Western readers to the damage well-meaning thoughtlessness routinely causes.

Watching the violence that was unfolding in Kenya, the aid world's self-justifying incantation kept echoing in my ears. 'The overall trajectory is up, the line of travel is forward,' development officials had repeatedly told me to justify support for Kibaki's government. Oh, really? I thought bitterly as I watched Kisumu's evisceration. The public's deepening sense of unfairness, caused by the very abuses the donors were determined to ignore, eventually bloodied the machetes in Kenya. Kenya's foreign partners failed to grasp that a system of rule based on the 'Our Turn to Eat' principle was explicitly designed to prevent the trickle-down upon which they counted for progress. The better Kenya's economy fared, the more unstable the country actually became, because public awareness of inequality – sociologists call the phenomenon 'invidious comparison' – deepened a notch.

It was a poor bet for the donors to make, for nothing sabotages development programmes more dramatically than violence. Decades of work on school-building, AIDS prevention and gender-awareness-raising are wiped out in a moment when the first *shamba* goes up in flames and its terrified family hits the road. Convinced they grasped the big picture, the donors somehow managed to miss the approaching near-collapse of an African state.

A country's economic prospects cannot be disentangled from its politics. The Anglo Leasing scandal's basic ingredients – ethnic arrogance and the growing sense of exclusion of a balked generation – provided the building blocks for the crisis that followed. The Mount Kenya Mafia's assumption that it could indefinitely occupy State House, in defiance of voters' wishes, marked the nadir of a journey that began with a score of too-good-to-refuse procurement deals. As South African analyst Moeletsi Mbeki told me: 'What greater corruption could there be than stealing an election?'

These are not merely Kenyan issues. Ethno-nationalism is emerging as Africa's most toxic problem, challenging the continent's very post-colonial structure. It has fractured Somalia, divided Ivory Coast, and is tearing at the fabric of Sudan and the Democratic Republic of Congo. As for the Western tendency to turn a blind eye to blatant graft and routine human rights abuse in the eagerness to save 'the poorest of the poor', it is a feature of donor relations across the continent. Worried Westerners, who so often seem to fall prey to a benign form of megalomania when it comes to Africa, would do well to accept that salvation is simply not theirs to bestow. They should be more modest, more knowing, and less naïve. They owe it not only to the Western taxpayers who make development organisations' largesse possible, but to the Africans whose destinies they attempt to alter.

In the time taken to write this book, things have not gone smoothly for the other high-profile players attempting to challenge official sleaze in Africa. In Sierra Leone, the head of the donor-funded Anti-Corruption Commission, Val Collier, was sacked in late 2005 after saying most MPs were putting their financial interests as government contractors before the nation's. In Nigeria, reformist minister Ngozi Okonjo-Iweala, who battled to change her country's ranking as the world's most corrupt, resigned in 2006 after being removed as head of the Economic Intelligence team. Nuhu Ribadu, chairman of the Economic and Financial Crimes Commission, was reassigned in December 2007 and fled the country after bullets were fired at his car. Following in John Githongo's footsteps, he has taken up the offer of a senior associate's post at St Antony's. In South Africa, Leonard McCarthy, head of the Scorpions, an FBI-style crime-fighting unit probing alleged police links with organised crime, learnt in February 2008 that the force was to be disbanded.

In theory, these are the individuals donors should be supporting. In practice, they often do the opposite. 'If you pump money into a system where there is leakage, you are effectively rewarding leakage and disincentivising those trying to stop it,' says Paul Collier. 'Change in Africa can only come from Africans, who are fighting against terrible odds. On the whole, they fail. They end up in exile, or come to a sticky end. If you don't, as a donor, support people like John, you are counteracting their fight for change.'

If they only set foot on the continent, idealistic Westerners would be astonished to hear how often, and how fiercely, politically engaged Africans – many of whom have featured in these pages – call for aid to be cut, conditionalities sharpened. Kenyan journalist Kwamchetsi Makokha is not alone in detecting an incipient racism, rather than altruism, in our lack of discrimination. 'Fundamentally the West doesn't care enough about Africa to pay too much attention to how its money is spent. It wants to be seen to do the right thing, and that's as far as the interest goes.' By subjecting donor budgets to unprecedented scrutiny, the global recession may, ironically, succeed where any number of sceptical reports on aid have failed, making it impossible for Africa's foreign backers to maintain their Pollyanna perspectives.

One of the many lessons of John Githongo's story is that the key to fighting graft in Africa does not lie in fresh legislation or new institutions. To use the seemingly counter-intuitive phrase of Danny Kaufmann, expert on sleaze: 'You don't fight corruption by fighting corruption.' Most African states already have the gamut of tools required to do the job. A Prevention of Corruption Act has actually been on the Kenyan statute book since 1956. 'You don't need any more bodies, you don't need any more laws, you just need good people and the will,' says Hussein Were. In Kenya, as in many other countries, the KACC is part of the grand corrupters' game, providing them with another bureaucratic wall behind which to shield, another scapegoat to blame for lack of progress.[44] Rather than dreaming up sexy-sounding short cuts, donors should be pouring their money into the boring old institutions African regimes have deliberately starved of cash over the years: the police force, judicial system and civil service.

Yet even today in Kenya, donors will get excited about the imminent approval of an anti-money-laundering Bill, or suggest that if only the KACC could be given prosecutorial powers – that old mantra – everything would be different. Things would be different, all right. With the likes of Justice Aaron Ringera at its head, a KACC with the power to prosecute could serve as an even more formidable obstacle in the fight against graft.

Donors would do better focusing on removing the beam from their own eye, by targeting the Western companies, lawyers' chambers and banks which make it possible for crooked African leaders to spirit hundreds of millions of dollars out of the continent each year. Deepak Kamani and his brother Rashmi, suspected of involvement in a dozen Anglo Leasing contracts, regard Britain as a second home. *Guardian* journalist David Pallister has tracked the Kamanis' business links with a Scottish arms dealer who funnelled Kenyan funds through accounts in Jersey and Guernsey.[45] His wife signed the first Anglo Leasing contract, and businessmen in Liverpool, Cambridge and Daventry played facilitating roles. These things are happening on British territory, with the active cooperation of seemingly respectable British companies.

In July 2007, a full two and a half years after John Githongo's departure from Kenya, Britain's Serious Fraud Office (SFO) and the City of London Police finally launched an investigation into British links with Anglo Leasing. Investigators took days' worth of testimony from John, and a dozen premises in London, Cambridgeshire, Liverpool and Scotland were raided. If the time taken by the SFO to interest itself was telling, so was the outcome. In February 2009 the SFO announced it was dropping its probe, citing a lack of cooperation by the Kenyan government. The SFO blamed Attorney General Amos Wako, Wako lambasted the SFO, Ringera blamed both Wako and Kenya's high court. When it comes to legalistic foot-dragging, the Kenyans have been able to imbibe lessons from the best. Their old colonial master still pays no more than lip-service to the anti-corruption cause. Since the OECD anti-bribery convention came into force in 2002, Britain has brought only four cases against domestic companies for bribery abroad, while the US has brought 120 and Germany 110.[46]

When former Kenyan justice minister Kiraitu Murungi described Anglo Leasing as 'the scandal that never was', he could not have been more wrong. Kenyans are still paying for Anglo Leasing, with the outgoings for several of the suspect contracts registering on the yearly budget. The Mars Kenya group has highlighted the ticking time bomb represented by the billions of shillings in irrevocable promissory notes issued in payment, whose final destination remains unclear. When former finance minister Amos Kimunya assured parliament in May 2007 that none had been issued for a digital communications network never supplied to the administrative police, he was immediately made to look ridiculous by Maoka Maore, the MP who first exposed Anglo Leasing, waving a fistful of such notes worth 49.7 million euros. Despite being deemed by Kenya's auditor general to have no legal existence, Anglo Leasing's 'ghost' firms are actually suing the Republic of Kenya for breach of contract. A responsible government would seize upon any court case as the perfect opportunity to flush out the players behind this network of fictional suppliers and phantom financiers. Instead, a colluding state is quietly reaching out-of-court settlements with the litigants. Although supposedly listed as 'wanted' by Kenyan police, in May 2008 Deepak Kamani returned to Nairobi, where he met with top officials. He is said to be considering a permanent return.[47]

What of John Githongo himself? As anyone who has watched old Westerns knows, the quest for retribution is doomed ultimately to disappoint; the lone avenger, no matter how righteous his cause, always ends up thwarted at some fundamental existential level, his vindication acid in his mouth. Being proved right is never enough.

Still single, he works out regularly and retains the silhouette of the jock who kicks sand in the wimp's eyes, but his stubble is now sprinkled with white. His multiplicity of talents have proved as much a curse as a blessing. By the time a man reaches his mid-forties, certain decisions need to be made. Unsure whether he wanted to be a politician, academic, anti-corruption campaigner or NGO director, John for years refused to make those decisions, trying, typically, to do it all.

By August 2008, however, one thing had become clear: he could no longer bear to remain in exile. Determined to be back in Kenya ahead of the storm he felt sure was approaching, he took up an invitation extended by the new coalition government and made his first visit in more than three years. He flew in to Jomo Kenyatta airport braced for possible arrest, his medication tucked into a jacket pocket in anticipation of a stretch in the cells, two private bodyguards at his side and another at the wheel of a waiting car. The moment he registered that there were no intelligence agents at the airport, only a phalanx of grinning family members and flashbulb-popping cameramen, he knew his enemies had dropped any notion of pressing treason charges.

In February 2009 he checked out of St Antony's for good, moving back to a tense Nairobi, where a fresh spate of top-level scandals, some involving names familiar from Anglo Leasing, had fuelled a mood of public fury. He headed straight for areas traumatised by the election violence and set about the slow task of grassroots mobilisation, encouraging communities to become more aware of their rights, better at lobbying for change. John pins his faith – that same passionate, all-or-nothing faith he once pinned on Kibaki – on the young, dispossessed *wananchi* who, he wants to believe, have grown wise to the ethnic string-pulling of their leaders. A sense of class solidarity, he argues, is forming in a group which finally grasps how cynically it has been manipulated.

These days he skirts the question – posed whenever his name crops up – of whether he intends to enter Kenyan politics, a move that would fulfil a long-nurtured sense of personal predestination. He has spoken in the past of running as an MP in Mathare slum, a St Francis-style itinerary for the spoilt Karen boy. But many believe his status as a Kenyan Brahmin disqualifies him from this course. 'His lineage is too privileged, too patrician,' says David Ndii. 'Ultimately, politics in Kenya is representative. Who does John represent? He doesn't have a Kenyan CV.' My own scepticism is based on different factors. Over-intellectual individuals who grind away at decision-making and shy from personal confrontation – and John is not a man to look anyone in the eye and say 'No' – may make brilliant deputies and priceless seconds-in-command, but rarely good prime ministers or presidents. Those roles need a capacity for judgement and sudden action which he lacks. Peppering his conversation with words like 'lustrations', 'parameters' and 'exponential curves', John does not possess the common touch. It may well be that he has already made his most important contribution to Kenyan history. But on a continent hungry for heroes, it's difficult to see his admirers allowing him to take the backroom job that would actually suit him best. A group of supporters has quietly registered a political party, no more than a name at present, which quietly awaits the Big Man's kiss of life.

John's permanent return coincided with the publication of this book, which was immediately deemed 'too hot to handle' by Kenya's Asian booksellers, scarred by memories of Moi-era libel suits. Their refusal to stock the book is a sure indication of how jittery Nairobi remains below a show of frenetic normality. The booksellers' boycott ensured that *It's Our Turn to Eat* briefly became the most pirated book in Kenyan history, with a bootleg PDF file of the manuscript circulating on the internet. But it did little to prevent the book reaching its audience. Backed by a Western development agency and an anonymous philanthropist, Kenya's churches, some NGOs and a media group formed an unlikely coalition dedicated to getting the book out to ordinary Kenyans, giving it to radio callers, debating it at public meetings and selling it at Nairobi's traffic lights.

John has been served with a summons by his old nemesis Chris Murungaru, who is suing for defamation. But the fury of those whose money-making schemes he sabotaged remains his main concern. The March 2009 assassinations, a stone's throw from State House, of two human rights campaigners who had dared publicise more than a thousand extrajudicial executions by the police were a sinister reminder that those who challenge the Kenyan establishment often court a death sentence. John takes extreme care, never moving alone. His life will never be entirely free of the fear of the assassin's bullet. 'John's a tragic figure, in a way,' a Kenyan television executive told me, 'because those who admire him cannot protect him, and those who hate him have very long memories.'

A polariser of opinions, a man destined to be either adored or reviled, John knows his path will never be ordinary. Sometimes he fantasises about a conversation with his old boss, in which he tells Kibaki to his face: 'You never believed in it, did you?' But he knows it will never occur, and that what he did, in the eyes of even former friends, will always verge on the unforgivable, exposing a capacity for cool calculation they struggle to accept. 'If your mother is naked, you throw a blanket on her, you don't call the neighbours round to have a look,' a Kikuyu acquaintance told me on my last trip to Kenya, capturing that sense of distaste. 'The hardest part,' acknowledges John, 'has been coming to terms with the betrayal of my tribe, my class.'

American writer Samantha Power coined the term 'upstander' – as opposed to 'bystander' – for those who decide where to draw a line and then refuse to cross it. Most Africans will face, at some point in their lives, a John Githongo moment of their own when they must choose whether to challenge The Way Things Have Always Been Done

Around Here, or to acquiesce. The dilemma goes to the heart of the puzzle of what it is to be a modern young African, forcing each individual to decide who he is and where his allegiance lies. The student who told me, 'I want to live in a country where it doesn't matter who my father is and where my family comes from,' was speaking for many Kenyans, but not all.

A crystallisation of that wrenching quandary, John's story attests to the qualities so often required by those who lead the way in breaking with the past: an intransigence verging on egomania, a literalism that is almost a form of foolishness. 'The worst thing I've been called is "naïve",' says John. 'I accept that. Only a naïve person would take an anti-corruption job after twenty-four years of systemic corruption. But that's what you need. I went in naïve and I want to stay that way.'

When I think of John's magnificently foolhardy attempt to bring down a system, an image comes to mind. It is of the carrion-eating marabou storks that nest in the scruffy thorn trees clustered around Nairobi's Nyayo sports stadium, on the airport road. '*Nyayo*' means 'footsteps' in Kiswahili, and the incoming Moi adopted it as his motif in order to reassure nervous Kenyans he would not deviate from Kenyatta's path. When it came to grand corruption this was certainly the case, and it has continued to be true under Kibaki. As sleaze has flourished, Kenya's poverty has deepened and the slums near the stadium have spread, along with the mounds of rotting rubbish on which the storks depend. Politicians, foreign aid officials and World Bank spokesmen continue to talk up Kenya's prospects, but each year I notice that there are more of these bald-headed scavengers perched above the honking *matatus* and diesel-snorting exhausts, a quiet, tangible sign that inequality is growing, not diminishing. Occasionally one will launch itself into flight, dark wings whooshing in a heart-stopping Nosferatu moment, fleshy pink throat sac swinging obscenely. But mostly they stand like frozen sentinels on the uppermost branches of the acacia, shoulders hunched, heads buried, bearing grim witness to the ingrained cynicism of an increasingly unjust society.

The John Githongos of this world have their work cut out.

Colin Bruce is no longer World Bank country director for Kenya. He left soon after the election crisis, in a swirl of controversy. An internal World Bank memo, leaked to the Financial Times, *revealed that even as one election monitoring team after another ruled the polls deeply flawed, Bruce was assuring his bosses in Washington that his landlord was the legitimate president of Kenya. A botched attempt to broker a deal between opposition and government – rejected by both sides – triggered criticism from across the political spectrum. His career does not appear to have suffered permanent damage, however. In June 2008 he was appointed Director for Strategy and Operations in the Africa Region, overseeing World Bank funding for the entire continent. Shortly before his reassignment, the World Bank's Compliance Advisor Ombudsman issued a memo to all Africa staff which named no names but had clearly been prompted by circumstances in Nairobi. 'I would like to remind staff of the need to avoid situations where potential conflicts of interest may arise (or be perceived to arise), or where the Bank may be exposed to possible reputational risk,' it said. 'With that in mind, the Region will not support international staff who are on assignment in country offices in entering into rental contracts with a counterparty who is (or is closely related to) a member of a client government, or has a well known affiliation with any major political party in a client country.'*

Retirement has not softened Sir Edward Clay. Ever alert to institutional hypocrisy, he follows African politics closely and is a vigorous writer of letters to British newspapers. 'Like diabetes in people of my age, it is perhaps a case of late-flowering or late-onset iconoclasm.' His frankness has prevented him building a post-ambassadorial career in the civil service. He was offered a job on the Foreign Office's Chevening Scholarships programme, which gives grants to overseas students, on condition he signed a gagging order which would have made it impossible to express opinions on current affairs. The offer was withdrawn after he publicly criticised the dropping of the inquiry into BAE, leaving him acutely aware of what he calls 'a symmetry of ostracism' on the part of both the Kenyan and the British governments. In the course of a BBC Hard Talk *discussion with justice minister Martha Karua in 2007, he discovered he had been made* persona non grata *in Kenya. During that exchange, Karua referred to his supposed real estate holdings in Kenya, often cited by the NARC elite as the reason for his interest in the country. Sir Edward owns no property in Kenya. He remains popular with ordinary Kenyans, and his remarks are often splashed across the local papers. A* matatu *baptised 'Edward Clay' has been spotted plying the roads of western Kenya.*

Caroline Mutoko is still at Kiss FM. Her morning broadcasts during the election crisis, urging listeners to refrain from violence, take care of their neighbours and return to work, were regarded by many listeners as providing the moral leadership the nation failed to receive from its politicians.

Conrad Akunga came close to packing in his Mzalendo website. 'I almost gave up. I'd spent five years being told, "Your vote is your voice and it will be heard." Then you queue for three kilometres in the sun, and all for what?' But he rallied, and is working on technical improvements. 'As long as those guys are there taking salaries, they are answerable to me. I'm their boss, not the other way round.'

Hussein Were was held up by armed men while entering his compound in August 2007. He escaped his attackers, whose motives remain unclear, by driving through a security barrier at top speed, shaking off a gunman clinging to his side mirror.

Were's former boss, Justice Aaron Ringera, was reappointed head of the Kenya Anti-Corruption Commission in August 2009 by President Kibaki, breaching legislation requiring the prior approval of parliament and the KACC board. The judge was forced to step down the following month after an unprecedented showdown between the president and incensed MPs.

On a sunny afternoon on 22 December 2007, John Githongo's former fiancée, Mary Muthumbi, married land surveyor James Ndirangu at a Catholic centre on a green hill in Dagoretti. The bride wore an ivory-coloured silk dress and her hair up. A matron of honour, four bridesmaids, five little flower girls and one embarrassed page hovered in attendance.

KEY CHARACTERS

Moody Awori – Vice-president in first NARC government
Colin Bruce – World Bank country director 2005–08
Edward Clay – British High Commissioner to Kenya 2001–05
Makhtar Diop – World Bank country director 2001–05
Alfred Getonga – Presidential aide in first NARC government
Dan Gikonyo – Physician to President Mwai Kibaki
Lisa Karanja – Former colleague of John Githongo, now runs TI-Kenya
Jomo Kenyatta – First president of independent Kenya, died 1978
Uhuru Kenyatta – Jomo Kenyatta's politician son
Mwai Kibaki – President of Kenya
Amos Kimunya – Lands minister in first NARC government, later appointed finance minister, suspended over controversial sale of Grand Regency hotel
Daniel arap Moi – Kenya's second president, in power for twenty-four years, retired in 2002
David Munyakei – Central Bank clerk who blew the whistle on Goldenberg scandal
Chris Murungaru – Minister for internal security in first NARC government. Dropped from cabinet in 2005
Kiraitu Murungi – Justice minister in first NARC government, he 'stepped aside' after leaking of Githongo dossier. Energy minister in current coalition
Francis Muthaura – Head of Kenya's civil service
David Mwiraria – Finance minister in first NARC government, he 'stepped aside' after leaking of Githongo dossier
Raila Odinga – former Kibaki ally, went into opposition in 2005. Probable winner of 2007 elections. Prime minister in current coalition government
Justice Aaron Ringera – Head of the Kenya Anti-Corruption Commission
George Saitoti – Education minister in first NARC government, now minister for internal security

GLOSSARY

askari – Kiswahili, 'guard'.

boda boda – Kiswahili, 'taxi bike'.

DfID – Britain's Department for International Development, set up in 1997 by Tony Blair's New Labour government to tackle poverty in the developing world. Before that, the Foreign Office was responsible for aid.

GEMA – Gikuyu, Embu, Meru Association. Set up in 1971 to promote the interests of the Mount Kenya ethnic communities, it won Kenyatta's disapproval and was disbanded in 1980.

GSU – General Service Unit. Elite Kenyan paramilitary force, responsible for internal security, made up of highly trained police officers and special forces. Often deployed to quell civil unrest.

KACA – Kenya Anti-Corruption Authority. Created by president Moi at the donors' insistence, it was ruled unconstitutional by the Kenyan High Court in January 2001.

KACC – Kenya Anti-Corruption Commission. KACA's more robust replacement, was formed in May 2003. Headed, until September 2009, by Justice Aaron Ringera.

KANU – Kenya African National Union. Political party which ruled for nearly forty years. Largely made up of Luos and Kikuyus, KANU initially faced a challenge from KADU, a federalist party set up by the smaller tribes. In 1982, Kenya became a one-party state, a situation which lasted until December 1991, when the constitution was altered at the donors' insistence.

kijana – Kiswahili, 'boy', 'my lad'.

kitu kidogo – Kiswahili, 'a little something', i.e. a bribe.

KNCHR – Kenya National Commission on Human Rights: independent watchdog set up in 2002 by an act of parliament.

matatu – Kiswahili, 'taxi bus'.

mzee (plural *wazee*) – Kiswahili, 'old man'.

mzungu (plural *wazungu*) – Kiswahili, 'white person'.

NARC – National Rainbow Coalition. Coalition of parties which toppled president Daniel arap Moi in the 2002 elections. It originally included Luo leader Raila Odinga's Liberal Democratic Party and fielded Mwai Kibaki as presidential candidate.

ODM – Orange Democratic Movement, opposition party set up in 2005 by Raila Odinga after spearheading a successful campaign against a new constitution proposed by Kibaki.

shamba – Kiswahili, 'farm'.

wabenzi – Slang, a member of the new African ruling class.

wananchi (singular *mwananchi*) – Kiswahili, 'citizens', 'the people'.

wazungu – See *mzungu*.

ACKNOWLEDGEMENTS

This book is based on a score of interviews and conversations with John Githongo in London, Oxford and Guatemala City between February 2005 and December 2008.

President Mwai Kwibaki and serving or former ministers David Mwiraria, Kiraitu Murungi and Chris Murungaru were all asked for interviews but none of them took up the offer.

Particular thanks go to film-maker Peter Chappell, for so generously sharing his insights and patient moral support as we both pursued our tantalising prey.

Andrew Hill was a constant sounding board. Kwamchetsi Makokha, Juliette Towhidi, Mutiga Murithi and my father, Professor Oliver Wrong, were all rigorous readers of the draft manuscript, alerting me to inconsistencies, errors and omissions. Mark Ashurst was a steady source of advice and encouragement.

In Nairobi, I owe Andrea Bohnstedt, Mahmud Abdulla, Kiki Channa, Eliot Masters, Susan Linnee, Ilona Eveleens, Koert Lindyer, Marina Rini, Massimo Alberizzi and Andrew England heartfelt thanks for beds in their homes, places at their tables and seats in their cars.

I thank Professor John Lonsdale for sharing his historical expertise, Professor Mutu wa Gethoi for his insights into Kikuyu culture, Professor Paul Collier for his economic analysis, Tom Wolf for his understanding of Kenyan politics and Sheetal Kapila and Pheroze Nowrojee for their legal advice. I'm specially indebted to David Cornwell, better known as John le Carré, for his tactful recommendations and wise counsel.

Apart from the Kenyans consulted and quoted in this book's pages – many of whom will certainly disagree with my views – I owe thanks to John Kamau, John Ngumi, Mary Mwirindia, Eric Wainaina, Benson Riungu, Waithaka Waihenya and Chand Bahal, owner of Bookstop, the best bookshop in Nairobi.

The Ford Foundation, whose Kenyan branch has played a pioneering role in the country's fight against corruption, provided funding that made the three years I took to write the book less of a financial strain than they would otherwise have been. Miles Morland, patron of African arts, cantered to the rescue when exasperation loomed.

The National Council of Churches of Kenya, Catholic Justice and Peace Commission, PEN Kenya, Peacenet Kenya, Open Society, the *Nairobi Star*, Kiss FM, East FM and Classic FM all took my breath away with the flair and determination they showed distributing my book in defiance of the booksellers' boycott, an unprecedented initiative pulled together by Galeeb Kachra and John Langlois at USAID in Nairobi.

Julian Harty devoted far too much of his precious time to tackling my computer problems. Michael Holman nudged me to realise my original idea.

But the book would never have seen the light of day had it not been for my commissioning editor, Mitzi Angel, who sadly moved on to a new publisher before the manuscript was complete, but did so much to improve it before she left.

Finally, my biggest practical debt remains to Joseph Githinji, whose thirty-year-old red-velvet-lined Volvo and illuminating conversation have kept me cheerful and on the road during my time in Nairobi.

MICHELA WRONG

London, September 2009

NOTES

1 Jomo Kenyatta was not alone in regarding the word 'Kikuyu' as a European deformation of the more correct 'Gikuyu'. However, 'Kikuyu' so predominates in both the Kenyan media and academic writing that using 'Gikuyu' – now widely reserved for the language – to refer to the community strikes me as pretentious.

2 'Kenya's Chance for a New Beginning', *Financial Times*, 30 December 2002.

3 Chinua Achebe, *The Trouble with Nigeria*, Fourth Dimension, 1983.

4 Bruce Berman and John Lonsdale, *Unhappy Valley: Conflict in Kenya and Africa, Book Two*, James Currey Ltd, 1992.

5 Wangari Muta Maathai, *Unbowed: One Woman's Story*, Heinemann, 2004.

6 Michael Blundell, *A Love Affair with the Sun: A Memoir of Seventy Years in Kenya*, Kenway Publications, 1994.

7 Gerard Prunier, 'Kenya: Histories of Hidden War', www.opendemocracy.net, 29 February 2008.

8 Karuti Kanyinga, 'When Figures Count: Governance, Institutions and Inequality in Kenya', Society for International Development (SID).

9 'Ministers' Home Areas Get Lion's Share of Roads Cash', *Daily Nation*, 20 July 2006.

10 Jean-François Bayart, Stephen Ellis and Beatrice Hibou, *The Criminalisation of the State in Africa*, African Issues Series, James Currey Ltd, 1999, p.103.

11 Transparency International – Kenya, 'Kenya Bribery Index 2001', www.tikenya.org.

12 M. Brockerhoff and P. Hewitt, 'Ethnicity and Child Mortality in Sub-Saharan Africa', Population Control, New York, 1998.

13 'Public servants would have engaged in business covertly anyway if they had not been allowed to do so by regulation': Duncan Ndegwa, *Walking in Kenyatta Struggles: My Story*, Kenya Leadership Institute, 2006, p.498.

14 By 1998, Kenya's stock of pending bills stood at an estimated 22 billion shillings. As Transparency International reported in its 2002 report 'Public Resources, Private Purposes', 'pending bills remain a slow time bomb that will explode in the face of Kenyan taxpayers'. Another useful paper on the various scams used throughout Kenyan history is Gladwell Otieno, 'The NARC's Anti-Corruption Drive in Kenya: Somewhere Over the Rainbow?', *African Security Review*, Vol. 14, No. 4, 2005.

15 Peter Warutere, 'The Goldenberg Conspiracy: The Game of Paper, Gold, Money and Power', Institute for Security Studies, Paper 117, September 2005; 'Report of the Judicial Commission of Inquiry into the Goldenberg Affair', Chairman Hon. Mr Justice S.E.O. Bosire, October 2005.

16 'A Survey of Seven Years of Waste', Centre for Governance and Development, Nairobi, February 2001.

17 'If you take a skunk home as a pet willingly, it's yours, together with its disturbing fragrance. It's disingenuous of you to blame the person you took it from for the smell and it is equally dishonest for the person who gave it to you to point at you and scream that these days you smell': John Githongo, *East African Standard*, 30 September 2006.

18 Jomo Kenyatta, *Facing Mount Kenya*, East African Educational Publishers Ltd, 1938. Other key publications on pre-colonial Kikuyu society are L.S.B. Leakey, *Mau Mau and the Kikuyu*, Methuen and Co. Ltd, 1952; Fr C. Cagnolo, *The Agikuyu: Their Customs, Traditions and Folklore*, 1933, revised edition 2006.

19 Colonel Richard Meinertzhagen, *Kenya Diary 1902–1906*, Oliver & Boyd, 1957.

20 Godfrey Muriuki, *A History of the Kikuyu 1500–1900*, Oxford University Press, 1974.

21 *Time* magazine, 30 March 1953.

22 David Anderson, *Histories of the Hanged: Britain's Dirty War in Kenya and the End of the Empire*, Weidenfeld & Nicolson, 2005.

23 E.S. Atieno Odhiambo and John Lonsdale, *Mau Mau and Nationhood*, James Currey Ltd, 2003, p.46.

24 Anderson, *Histories of the Hanged*, op. cit., Caroline Elkins, *Britain's Gulag: The Brutal End of Empire in Kenya*, Jonathan Cape, 2005, and Lotte Hughes, *Moving the Maasai*, Palgrave Macmillan, 2006, are the most notable examples.

25 Blundell, *A Love Affair with the Sun*, op. cit.

26 Society for International Development, 'Pulling Apart. Facts and Figures on Inequality in Kenya', October 2004.
27 Kenya National Audit Office, 'Special Audit Report of the Controller and Auditor-General on Financing, Procurement and Implementation of Security Related Projects', April 2006.
28 Mars Group Kenya GAP Report No. 2, 'Illegally Binding: The Missing Anglo-Leasing Scandal Promissory Notes', www.marsgroupkenya.org.
29 NARC's Assistant Justice Minister Njeru Githae, 'New Plan to Recover the Looted Billions', *Daily Nation*, 16 July 2003.
30 Paul Collier, *The Bottom Billion*, Oxford University Press, 2007, p.109.
31 Stephen Brown, 'Authoritarian Leaders and Multiparty Elections in Africa: How Foreign Donors Help to Keep Kenya's Daniel arap Moi in Power', *Third World Quarterly*, Vol. 22, No. 5, 2001, pp.725–39.
32 Bronwen Maddox, 'Cracks Under Surface of the £5bn Labour Mission for World's Poor', *The Times*, 19 March 2007.
33 Department of Institutional Integrity, World Bank, 'Kenya Detailed Implementation Review Report', 10 January 2007. This strictly confidential report was published on the website of the *Wall Street Journal* ('Kenya and the World Bank') on 6 March 2008.
34 Richard Beeston and Xan Rice, 'Regime is Told to Clean up its Act or Pay Price', 'Watchdog Muzzled by his Master Prepares to Bark', *The Times*, 21 January 2006.
35 'Exclusive: The Anglo Leasing Truth', *Daily Nation*, 22 January 2006.
36 The Kenyan government commissioned an inquiry into the Artur brothers saga, dubbed the Kiruki Commission. In April 2007 the government announced that its findings would not be publicly released for reasons of 'national security'.
37 Public Accounts Committee, 'Report on Special Audit on Procurement of Passport Issuing Equipment by the Department of Immigration, Office of the Vice-president and Ministry of Home Affairs', Kenya National Assembly, March 2006.
38 UNDP, 5th Kenya Human Development Report, 2006.
39 British aid to Kenya rose from £24.9 million in 2002–03 to £64.2 million in 2005–06, according to DfID's annual reports. Despite talk of contracts with British companies being cancelled under Kibaki, UK exports to Kenya were £214.98 million in 2006, compared to £158.86 million in 2002, the last year of the Moi presidency, according to the UK Department of Trade and Industry. With £1.5 billion in investments, the UK remains by far the largest foreign investor in Kenya.

40 'Ethnicity and Violence in the 2007 Elections in Kenya', Afrobarometer Briefing Paper No. 48, February 2008; Roxana Gutierrez Romero, Mwangi S. Kimenyi and Stefan Dercon, 'The 2007 Elections, Post-Conflict Recovery and Coalition Government in Kenya', *Improving Institutions for Pro-Poor Growth*, 26 September 2008.

41 Well-researched reports on the 2007–08 election crisis include: 'Still Behaving Badly', Second Periodic Report of the Election-Monitoring Project, December 2007; Kenya National Commission on Human Rights; 'Kenya in Crisis', Africa Report No. 135, Crisis Group, 21 February 2008; and 'Ballots to Bullets: Organised Political Violence and Kenya's Crisis of Governance', *Human Rights Watch*, Vol. 20, No. 1 (A), March 2008.

42 David Anderson and Emma Lochery, 'Violence and Exodus in Kenya's Rift Valley, 2008: Predictable and Preventable?', *Journal of Eastern African Studies*, Vol. 2, No. 2, July 2008.

43 Britain's 2005 Commission for Africa report, 'Our Common Interest', makes this point in its 'Culture' chapter, quoting academics Stephen Ellis and Gerrie ter Haar: 'No more than anyone else do Africa and Africans have an authentic, unchanging culture that is transmitted from one generation to another, or ought to be.'

44 cf Bayart, Ellis and Hibou, *The Criminalisation of the State in Africa*, op. cit., p.101: 'Special anti-corruption units and commissions exist essentially to attack political and economic rivals, while at the same time placating aid donors.'

45 David Pallister, 'Fraud Office Inquiry into UK Links to Kenyan Cash and Arms Scandal', *Guardian*, 1 October 2007.

46 Transparency International, 'OECD Anti-Bribery Convention Progress Report 2009' http://www.transparency.org/publications/publications/conventions/oecd_report_2009

47 'Did Kamani Meet Kibaki?', *Nairobi Star*, 18 May 2008.

INDEX

(JG indicates John Githongo)

a'Nzeki, Archbishop Ndingi Mwana 221
Achebe, Chinua 42, 132
Africa Commission 206, 212
Akunga, Conrad Marc 153, 154, 155, 158, 334
Al Qaeda 78, 258–9, 277
Ali, Major General Hussein 256
Anderson, David 107, 232, 308
Anglo Leasing and Finance Company Ltd 129, 162, 206; British investigation into 328; contracts with Kenyan government 85, 164, 165, 166, 169, 173, 180; cost of Kenyan government contracts with 165–6, 169, 174, 210, 216; elections, money from deals used to fund NARC campaigns ('resource mobilisation') 215–16, 219–20, 243; foreign donors' reaction to scandal 260–4, 266, 267; forensic laboratory contract 164, 173; 'ghost firms' sue Kenyan government for breach of contract 328; JG investigates 79, 80–97, 135, 163–6, 233–45, 248–54, 268–9; JG releases details of investigation into 248–54; KANU and 77–9; Kibaki and 219, 220, 222, 234, 235–6, 238, 244–5, 251, 265–6, 268, 273; Maore reveals scandal 77–9, 85, 86; media coverage of scandal 240, 245, 248–54, 255–6, 269–70; ministers involved with 84–97, 119, 165, 166, 171–2, 173, 177, 179, 215–16, 217–20, 222, 223, 245, 250–1, 268; Moi and 165, 171–2; NARC and 79, 84–97, 118, 165, 166, 171–2, 173, 177, 179, 215–16, 217–20, 222, 233–6, 241, 242–3, 245, 250–4, 261, 268, 269–74; navy frigate contract 165, 180; passport printing and lamination contract 78–9, 84–7; payments to 164–5, 170–1, 173, 219, 268, 328; police Mahindra jeep contract 78, 84; classic procurement scam 168–72; 'Project Nexus' 165; shadowy nature of 171–2
Annan, Kofi 314, 315
Anti-Corruption and Economic Crimes Act 66, 285
Artur brothers 256–60
Asians, Kenyan 139, 146, 168, 172, 173, 270, 281, 303, 316
Awori, Moody 84, 85, 87, 213, 245, 250, 251, 268, 301

BAE Systems 276–7, 334
Bellamy, William 193–4, 222–3, 224, 250, 259
Benn, Hilary 267, 276
Biwott, Nicholas 301
Blair, Tony 54, 194, 200, 206, 212, 276
Bland, Simon 267
Blixen, Karen 8–9, 121
Blundell, Michael 50, 112
Bomas of Kenya 74, 241
Bono 205, 266
Bosire Commission 65, 268, 321
British High Commission, Nairobi 194, 195, 198, 202–3, 253, 268, 296
Brown, Stephen 188

Bruce, Colin 260, 263, 275, 278, 333
Bush, George W. 262, 263, 275, 277
Buwembo, Joachim 306

Cameron, James 109
Castle, Barbara 109
Central Bank, Kenya 62, 78, 93, 164–5, 259, 318, 319, 320, 322
Central Province, Kenya 52, 73, 99, 102, 108, 113, 114, 117, 127, 147–8, 243, 280, 282, 297, 298, 301, 304
Charterhouse Bank Ltd 259–60
China 209, 240
Christian Aid 160, 205
Churchill, Winston 105
CIA 82–3
Clay, Sir Edward 71, 72, 171, 183, 194–204, 210–15, 224, 225, 254, 259, 267, 277, 333–4
Collier, Professor Paul 31, 51, 185–6, 232, 326
Collier, Val 326
Cornwell, David (John le Carré) 23

Daily Nation 63, 68, 91, 119, 131, 202, 222, 247–51, 253, 282–3, 291, 298, 319
De La Rue 78
Delamere family 8, 105
Democratic Party 69, 118, 135, 175–6, 242
Democratic Republic of Congo 15, 83, 115, 124, 196, 325
Department for International Development, UK (DfID) 194, 204, 205–12, 225, 261–2, 267, 275, 276, 337
Diop, Makhtar 190–2, 199, 223–4, 260
Dorobo tribe 103, 105

EastAfrican 12, 141
East African Community 140, 307
East African Standard 41, 56–60, 248, 255–7, 259
Easterly, William 207
Eigen, Peter 141, 142
Eldoret 112, 114, 290, 307
Electoral Commission of Kenya (ECK) 298, 301, 306, 307, 315n
Eliot, Sir Charles 46
Embu 43, 73, 108, 216, 291, 307, 313
End of Poverty, The (Sachs) 192
Escrivá, Josemaría 33, 134
Ethiopia 9, 50, 115, 192, 277
European Union: Common Agricultural Policy 207; elections 2007, observers in Kenya for 298, 307; gives greater say to Kenyan Treasury on how aid is spent 286; JG's resignation, reaction to 223
Executive magazine 65, 138, 139, 141

Face Technologies 78
Facing Mount Kenya (Kenyatta) 103–4, 107
FBI 228, 326
Financial Times 11, 12, 23, 187, 189–90, 333
Fish, Dave 267
Ford Foundation 264
Foreign and Commonwealth Office, UK (FCO) 194, 196, 201, 203, 206, 225, 259, 334

G8 meeting, Gleneagles (2005) 206, 212, 262
Gado 63, 202
Geldof, Bob 205, 266
Gellhorn, Martha 8
General Service Unit (GSU) 3, 13, 34, 79, 187, 311, 337
Gethi, John 129
Getonga, Alfred 72, 85, 87, 91, 129, 177, 179, 215, 251
Gikonyo, Dr Dan 70–1, 175, 242
Gikuyu language 26, 74, 136, 222
Gikuyu, Embu, Meru Association (GEMA) 113, 126, 144, 176, 243, 337
Githongo, Ciru (JG's sister) 39, 125, 133
Githongo, Gitau (JG's brother)125, 131, 132, 138, 239–40
Githongo, Joe (JG's father): accountant 126–7; attempts to influence JG through 94–5, 118–19; Catholic faith 123, 142, 143; home 123; JG and 15–16, 94–5, 119, 130–1, 133, 137, 175, 238–9, 250, 265; Kenyatta regime, role in 118,

160; London, life in 125; loner 135–6; marriage 123; Mau Mau and 123–4, 125; Moi regime and 118, 160; railways, work on 126; social standing 135–6; Transparency International, role in birth of 14, 141; tribalism, loathing of 126–7, 142–3; UN, work for 124–5
Githongo, John (JG): African roots, disconnection from 143–4, 145, 229, 266–7, 286–7; Anglo Leasing, compiles dossier on 233–5, 240–1, 242, 243, 244–5, 248–54; Anglo Leasing, investigation into government contracts with 78–9, 80–97, 163–6, 233–5, 240–1, 242, 243, 244–5, 248–54; Anglo Leasing, releases details of investigation to media 248–54; author, relationship with 12–13, 14, 19, 20, 23–31, 74; author, stays with 19, 20, 23–31; birth 125, 160, 229; Britain, connection with 125, 160, 229, 266–7, 286–7; Catholic faith 27–8, 133–5, 142, 143, 159, 174, 175, 176, 237, 248; character 12, 13, 19, 27, 35–6, 37–40, 81, 132–3, 135, 161, 175–7, 216–17, 233, 235, 236, 264–5, 318, 323, 329–31; Chief of the Burning Spear 31–2, 218; childhood 127, 130–1; diaries 32, 135, 216, 234; elections 2002, role in 14; exile 227–60, 263–7, 271–8, 329; family home 121–3; family, effect of actions upon 175–6, 216, 238–40, 250, 265; father, relationship with 15–16, 17, 118–19, 130–1, 137, 145, 175, 238–9; foreign donors and 15, 38–9, 194, 222–3, 235, 236, 261–2, 263–4, 266, 267; girlfriend 28, 91, 216–17, 221, 250, 264–5, 335; high treason charge, possibility of 237, 251; informers 81–4, 86, 172, 180, 232, 244–5; journalist 12, 14, 138–41, 143; KACC, testifies before 254, 256, 268, 271, 272; Kibaki, relationship with 14, 15, 17–18, 34–5, 37, 66–8, 71, 86, 87, 90–1, 172–5, 179–82, 217, 219, 220, 222, 235–6, 238, 244–5, 265–6, 311–12; Kikuyu ethnicity 13, 97, 115, 117–19, 130, 142–4, 145, 159–62, 163–4, 243, 253, 287–8, 289; legacy of revelations 322–4, 327, 328; media, relationship with 14, 22, 92, 240, 245, 248–50, 251, 252, 253–4; Michael Holman, stays with 39, 31; Ministry of Justice, role in setting up of 66; Moi, investigates corruption under 14, 86; mother, relationship with 133–4, 175, 238–9; NARC and 14, 66–8, 79–87, 164–6, 171–82, 215–25, 229, 233–6, 241, 242–3, 244–5, 250–4, 268–74; Oxford University, stays at 31, 32, 228–33, 235, 242, 243, 264, 285, 290, 322, 329; PAC, testifies before 95, 237–8, 254, 257, 268; Permanent Secretary in Charge of Governance and Ethics, appointment as 12, 15–18, 17, 36–7, 118–19, 145; Permanent Secretary in Charge of Governance and Ethics, abortive attempt to demote from post as 18, 177–82, 215; Permanent Secretary in Charge of Governance and Ethics, office 34–6; Permanent Secretary in Charge of Governance and Ethics, resigns as 20, 21–3, 220–5; physique 12–13, 132; police, joins 137–8; public view of 165–6, 252–4, 286–90, 322; return to Kenya 329–30; revolt, reasons for ability to 158–62, 175–6; school 127–9, 133, 135, 136, 175; smear campaign against 91–2, 231, 254, 264; State House office 34–5, 164; studies in England 12; taping of ministers' conversations 21, 32, 88–90, 233–5, 251, 252, 253–4, 273–4, 289, 321–2; threats to 18, 26, 37, 87–8, 90–7, 176, 218–19, 222, 227–8, 229, 230–2, 250, 257–8, 274, 329–30; Transparency International and 14, 15, 141–2; travel within Africa 12; university in Wales 136, 143; World Bank Volcker panel, recruited by 263, 278
Githongo, Mary (JG's mother) 123–4, 125, 133–4, 238–9
Githongo, Mugo (JG's brother) 39, 126, 136, 138, 158, 178, 221, 239, 240, 274–5

Githongo & Company 126–7
Goldenberg scandal 62–3, 65, 86, 89, 139, 165–6, 168, 211, 216, 221, 228, 251, 268, 269, 284, 311, 319, 320, 321, 322, 324
Goldsmith, Lord 276
Grand Regency hotel 63, 322

Hall, Francis 104
Hannan, Lucy 312
Hemingway, Ernest 8
Hileman, Milena 162
HIV/AIDS 166, 171, 208, 302, 325
Holman, Michael 30, 31
Home Guard 108, 110, 115, 123, 142, 288
Howells, Kim 259
Human Rights Watch 35

Infotalent Ltd 173
International Monetary Fund (IMF) 60–1, 170, 184, 186, 261
Iraq 262, 275
Islamic extremism 78, 188

Kaggia, Bildad 50
Kaiser, Father 26
Kamani, Deepak 84, 173, 327, 329
Kamani, Rashmi 327
Kamba people 43, 48, 58–9, 102, 116, 129, 157, 163, 313
Kantai, Parselelo 324
KANU party 68, 77, 86, 262, 337; Anglo Leasing and 78, 171–2; elections 1997 188; elections 2002 2; elections 2007 215; independence and 109; Kikuyu and 113, 114; long grip on power 2 *see also* Moi, Daniel arap
Karanja, Lisa 35–6, 38, 92, 95–6, 178, 217, 289, 324
Karen, Nairobi 121–2, 134, 135, 175, 242, 330
Kariuki, J.M. 26, 279
Karua, Martha 271, 311, 334
Karume, Njenga 175
Kaufmann, Daniel 207–8, 263, 327
Keane, Fergal 251
Kalenjin people 42; Asian Kenyans, relationship with 172; character of 43, 44; elections 2007 and 290, 297, 308, 312, 316; ethnic violence in the early 1990s, role in 114, 140, 141; history of 49, 112, 113, 114, 140, 141; invention of 49; Kenyatta and 112, 113; Moi and 51, 52, 56, 57, 142, 284, 300

KENYA

AID AND DONORS, FOREIGN: Anglo Leasing, reaction to scandal of 192, 260–4, 266, 267; anti-corruption drive 262–4, 266–7, 275–8; Cold War 183, 185; Consultative Group meeting (April 2005) 223; corruption, general approach to 8, 9–10, 14, 38–9, 51, 53–5, 60–4, 184–225, 260–4, 266–7, 275–8, 324–5, 327–8; DfID *see* Department for International Development, UK; Gleneagles (2008) 206, 212, 262; High Commissioner Clay raises corruption as an issue 193–204; history of aid to Kenya 183–204, 205–22; intimacy of relationship between donors and government ministers 190–3, 224; IMF *see* IMF; JG, relationship with 15, 38–9, 194, 197, 198, 222–3, 235, 236, 261–2, 263–4, 266, 267; JG, reaction to resignation of 220–5; KACA and 186; under Kenyatta 184; under Kibaki 11, 38–9, 197–204; legislation passed to appease 185–6; under Moi 8, 9–10, 14, 170, 184, 185–7; multi-party politics, enforces 51, 62; NARC regime and 170, 183, 210–11; as percentage of government spending 184; structural adjustment funding 160, 184; UK aid 187, 193–204, 205–12, 225, 261–2, 267, 275, 276–7, 337; vastly boosted aid, recent culture of 205–13, 262; War on Terror and 275–6, 277; World Bank *see* World Bank

COLONIAL RULE: Asian Kenyans under 172; clichés of 8–9; corruption under 60; economy under 11; education under 134; ethnic self-awareness, role in 45–50, 52; foreign aid and history of 184, 194, 202, 203; Karen under 122; Kikuyu under 104–10, 115; living

standards under 11; modern Kenyan attitudes towards 286–8; pre-eminent role of Kenya in Africa as a result of 8–9
CORRUPTION: Anglo Leasing and *see* Anglo Leasing and Finance Company Ltd; anti-corruption legislation 65–6, 283–4, 285–6, 327; banks and 61, 259–60, 270; bribes 11, 55, 337; challenges to pan-African 326–7; donor communities' approach towards 8, 9–10, 14, 38–9, 51, 53–5, 60–4, 184–225, 260–4, 266–7, 275–8, 324–5, 327–8; economy and 11, 60–1, 160–1; effect upon stability of country 258–9; elections 2002, NARC campaign on 2; history of 41–2, 50–3, 56–9, 60–3; Islamic terrorism and 258–9; JG fights *see* Githongo, John; judiciary 65, 268–9; Kenyan ambivalence towards 55–6, 166–8, 169, 287–8; Kibaki and *see* Kibaki, Mwai; *kitu kidogo* ('petty' corruption) 55, 337; land grabbing 61–2, 65; layers of Kenyan 41–2; legal system used to fight action against 268–9; Moi and *see* Moi, Daniel arap; NARC and *see* National Rainbow Coalition; NSIS investigations into 80–1; 'Our Turn to Eat' culture (patronage) and 8, 11, 35, 42, 44, 5–3, 56–60, 72–4, 112–14, 116–19, 143, 144, 145, 157–62, 272, 282, 290–2, 297–300, 325–6; 'pending bills' 61, 65; Permanent Secretary in Charge of Governance and Ethics, post created 11–12; public procurement and 65, 77, 78–97, 168–72, 200 *see also* Anglo Leasing and Finance Company Ltd; state loans and 61; 'straddling' 60; UN Convention Against Corruption, ratifies 65; Transparency International *see* Transparency International
CULTURE: character of Kenyan people 7, 67–8, 157; demography 150; education system 51, 127–9, 133, 135, 136, 175; history of 45–53; population, rise in 103, 147–8, 149, 195; recent shifts in 145–62; slums 2, 10, 11, 41–2, 149, 151, 157, 191, 203, 290, 303, 312, 330; tourism 9, 311, 317; urbanisation 148–9; youth 150–5
ECONOMY: 160–1; Cold War 9–10; colonial, effect of 9, 10–11; corruption, patronage ('Our Turn to Eat' culture) and 11, 60–1, 114, 160, 184–5; ethnicity, role in 282, 290, 299–300; under Kenyatta 184; under Kibaki 184, 220, 261, 279–82, 290, 292–3, 299–300; under Moi 2, 8, 9–10, 159, 171, 184, 261, 280, 290; most advanced in region 9; NARC and 184, 220, 261, 279–82, 290, 292–3, 299–300; post-election violence 2007, effect upon 311; rich and poor, wealth gap between 11, 279–80, 282, 290, 299–300; state sector shrivels 1980s 160
ELECTIONS: 1992 general election 140; 1997 general election 188; 2002 general election 1–2, 14–15, 73, 135, 154–5, 193, 215, 237, 241, 320; 2004 general election 214; 2005, referendum on new constitution 241–4; 2007 general election 290, 295–316; post-election settlement 2007 314–16
ETHNIC GROUPS: 11, 60–1; colonial rule, effect upon 44–50, 60, 104–10, 115; economy and 60–3, 282; elections 2007 and 290, 291–2, 295–316; ethno-favouritism/elitism 8, 11, 44, 50–3, 56–9, 72–4, 86, 112–14, 116–19, 143, 144, 145, 157–62, 272, 282, 290–2, 297–300, 325; ethno-nationalism as Africa's most toxic problem 325; history of 45, 46–7, 48, 49–53, 60–3, 104–14; JG and family break mould of 97, 118–19, 121–44, 158–62, 163–4; JG, Mount Kenya Mafia appeal to ethnicity of 97, 118–19; Rwanda, effect upon Kenyan awareness of 140–1; stereotypes 42–4, 49–50; under Kenyatta 50–1, 52, 110–13; under Kibaki 51–3, 56–60, 97, 118–19, 290, 291–2, 295–316; under Moi 51, 52, 113–14; violence between ethnic groups, early 1990s 114, 140,

141; youth outlook on 150–8 *see also under individual ethnic group name*
MEDIA: 7, 12, 131–2; Anglo Leasing deals, coverage of 245, 248–54, 255–6, 269–70; cartoonists 7, 10, 34, 63; corruption, coverage of 198, 240, 245, 247–54, 255–7, 269–70; JG and 12, 14, 22, 92–3, 245, 248–54, 269–70; KTN, commandos storm 255–6; Moi and 131–2; newspapers, independence of 131–2, 247–8; newspapers, quality of 12; radio 155–8, 334; television 132, 251, 255–6 *see also under individual newspaper title and radio and television station name*
POLICE: 3; Anglo Leasing supply jeeps for 78, 84; corruption within 6, 11, 77, 78, 84, 173; elections 2007, actions during 305, 312; digital communications contract 328; Infotalent Ltd security contract 173; shootings 192 *see also* General Service Unit (GSU)
POLITICS: constitution 8, 13, 73, 86, 113, 185, 241–4, 283; State House 12, 21, 22, 27, 32, 33, 34, 35–6, 38, 39, 65–6, 69, 70, 71, 74, 80, 81, 84, 89, 92, 93, 129, 160, 164, 172, 173, 174, 177, 179, 181, 183, 184, 202, 218, 219, 220, 222, 224, 232, 235, 238, 244, 249, 260, 270, 271, 283, 300, 304, 306, 325; multi-party regime, advent of 7, 51, 62, 79, 88, 113–14, 139, 140, 158, 193; NARC *see* National Rainbow Coalition; parliament 153; prime minister, NARC expected to install 8, 283
Kenya Anti-Corruption Authority (KACA) 186, 337
Kenya Anti-Corruption Commission (KACC) 20, 66, 84, 86, 87, 92–3, 173, 217–18, 222, 245, 250, 254, 256, 261, 269, 270, 271, 272, 273n, 274, 327
Kenya Broadcasting Corporation 132, 251
Kenya National Commission on Human Rights (KNCHR) 36, 80, 220, 291, 297, 324, 337
Kenya Revenue Authority 280
Kenyan High Commission, London 23, 30, 238, 254, 257, 271
Kenyan Television Network (KTN) 255–6
Kenyatta, Jomo 34, 36, 50–1, 73, 87, 103–4, 106, 107, 109, 110, 112, 113, 117, 118, 122, 125, 126, 159, 160, 184, 279, 286, 296, 297, 300, 332
Kenyatta, Uhuru 2, 10, 129, 175, 237
Keriri, Mateere 72, 175
Kettering, Merlyn 173
Khamala, Martin 138
Kiai, Maina 36, 291
Kiambu people 103, 105, 108, 116, 123, 136
Kibaki, David 129, 175
Kibaki, Lucy 190–1, 224
Kibaki, Mwai: Anglo Leasing dossier, reaction to release of 251, 268, 273; anti-corruption legislation 65–6, 283–4, 285; cabinet 7–8, 72–4, 80, 84–5, 241, 244, 245, 268, 273; car crash 2002 4, 34, 94, 181–2, 242; character 4–5, 8, 68–70, 71–3, 219; constitution and 8, 73, 86, 241–4, 283; corruption, attitudes towards 2, 5, 7, 12, 20, 65–6, 74–5, 77–8, 79–80, 81–97, 215, 219, 220, 222, 234, 235–6, 238, 244–5, 265–6, 282–4, 285; economy under 184, 220, 261, 279–82, 290, 292–3, 299–300; elections 2002 4, 73; elections 2007 284–5, 311–12, 314; ethnic favouritism under 7–8, 72–4, 85, 86; features of former era re-emerge under 282–4; foreign donors and 11, 38–9, 197–204; health 4, 34, 70–3, 80, 94, 181–2, 242; inauguration 2002 1–6, 8, 14, 65–6, 164, 283; inauguration 2007 306; JG, relationship with 14, 17–18, 22, 23, 34–5, 37, 67–8, 71, 86, 87, 90–1, 179–82, 219, 222, 234, 235–6, 238, 244–5, 265–6, 311–12; Moi and 5, 7, 68, 284–5; Mount Kenya Mafia and 72–4, 85, 172–3, 245; speeches 5, 72
Kibera slum, Nairobi 1, 149, 190
Kikuyu people: birth of tribe 101–3; burial of 147–8; Central Association

106; character of 43, 103–14, 115–19; colonial rule, under 105–10; country 98–103; culture 103–14; entitlement, sense of 117; foreign influences upon 131; history of 46, 47, 48, 101–14; JG and 13, 97, 115, 117–19, 130, 142–4, 145, 159–62, 172–5, 287–8, 289; jokes, enjoyment of 115–16; Kenyatta regime, benefit from 50–1, 52, 109–10, 111–13, 115, 159; Kibaki and 8, 7–8, 72–4, 85, 86; land ownership, importance of 104, 105, 112; Mau Mau and 107–12; under Moi 113–14, 159; multi-party politics, role in advent of 113–14; population 103; Reserve 106, 111; respect for elders 118; son and mother bond 159
Kikwete, Jakaya 314
Kimani, Martin 128, 159, 324
Kimathi, Dedan 111
Kimunya, Amos 242, 243, 250–1, 259, 285–6, 311
Kinyua, Joseph 219
Kiriamiti, John 316
Kisii tribe 307, 313
Kiss FM 155, 156, 334
Kisumu, Nyanza Province 282, 290, 298, 299, 300, 302–3, 306, 307, 325
Kiswahili language 26, 74, 79, 90, 150, 151, 203, 222, 332
Kivuitu, Samuel 298, 304, 307
Koinange, Jeff 129
Kroll 12, 86, 221, 284

Land and Freedom Armies 107 *see also* Mau Mau
land grabbing 61–2, 65
Langata Cemetery 145–8
Leakey, Philip 186
Leakey, Richard 18, 164, 186
Lone, Salim 284
Lonsdale, John 49, 111
Luhya people 43, 58–9, 116, 163, 307, 313
Luo people 43, 44, 49–50, 73, 116, 127, 129, 149, 216, 241, 289, 290, 291, 299, 302, 303, 304, 307, 308, 309, 310, 312, 313, 316

McCarthy, Leonard 326
MacKinnon, William 45, 46
Maasai 42, 43, 44, 46, 47, 48, 49, 103, 104, 105, 106, 112, 114, 124, 281, 320
Maasai Mara 53, 317
Maathai, Wangari 49–50, 175
Macharia, Edith 253
Magari, Joseph 84–5, 87. 93, 165
Maina, Wachira 42, 324
majimboism 296
Make Poverty History 205, 206, 266
Makokha, Kwamchetsi 131–2, 284, 297–8, 326
Makotsi, Pamela 256
Malik, A.H. 94
Mangu, Central Province 125, 131
Maore, Maoka 77, 78–9, 85, 86, 328
Margaryan, Artur 256–60
Mars Group Kenya 166, 324, 328
Mathare slum 290, 308, 309, 310, 330
Mati, Mwalimu 36, 67, 69, 79, 175, 217, 315, 324
Matiba, Raymond 129
Mau Mau 50–1, 107–12, 115, 123, 124, 125, 130, 140, 142, 159, 232, 287, 288
Mbeki, Moeletsi 325
Mboya, Tom 26, 299
Meinertzhagen, Captain Richard 104–5
Meru people 43, 73, 108, 216, 222, 291, 307, 313
Michuki, John 175, 256, 297, 311
Migiro, Dickson 254
Ministry of Justice, Kenya 40, 65, 179
Ministry of Finance, Kenya 218–19
Moi, Daniel arap 73, 178, 287, 301; anti-corruption bills 8; constitution and 13, 113, 185; corruption under 14, 39, 62, 66, 77, 80–1, 82, 86, 165, 171–2, 211, 214, 228, 270, 284–5, 286; *East African Standard*, involvement in 56–8, 256; economy under 2, 159, 160, 171, 184, 261, 290; elections 2007, role in 284–5, 313; foreign donors and 8, 9–10, 14, 170, 183, 185–6, 198–9; Goldenberg scandal, role in 62, 211; justice system under 65; Kalenjin ethnicity 51, 52, 56–8, 284, 300; Kibaki

and 4, 5, 7, 68, 284–5; Kikuyu, treatment under rule of 113–14, 117, 164; legacy 5; liberalisation of the airwaves mid-1990s 156; ministers, treatment of 67; multi-party rule forced upon 113–14, 139, 140; opponents, treatment of 7; retirement 2, 4, 34, 69, 70, 183, 197; Transparency International and 14
Moi, Gideon 129
Mount Kenya 43, 46–7, 73, 99, 100, 101, 103, 108, 117, 145, 271, 272, 302
Mount Kenya Mafia: birth of 73, 74; donor community and 222; elections 2007, role in 297, 311, 325; Goldenberg scandal, shelves 269; JG and 89, 96, 119, 143, 144, 161, 179, 203, 215, 230, 242, 273; Kibaki, level of influence over 85, 172–3; media, attacks 256; referendum on constitution 2005, involvement in 244; technology, weakness with 89–90, 230
Moyo, Jonathan 162
Muga, Wycliffe 39, 119, 143, 252, 253
Mugabe, Robert 162, 256, 287, 287n
Muhoho, George 15–16, 175, 297
Muite, Paul 148
Mukurwe Wa Nyagathanga 102
Mule, Harris 15–16, 31–2, 118
Muli, Koki 306–7
Mullei, Andrew 164–5
Mumbi 102, 115, 117, 144, 253
Munro, Bob 34
Munyakei, David 317–19, 320, 321–2, 323, 324
Murage, Stanley 179
Muranga 100, 101, 104, 108–9, 316
Murgor, Philip 177, 220
Murgor, Willie 112
Muriuki, Godfrey 105
Murungaru, Chris 72, 84, 85, 87, 88, 92, 177, 178, 179, 202, 212, 215, 220, 222, 223, 250, 268, 284, 301, 329
Murungi, Kiraitu 69, 72, 74, 86, 87, 88, 93, 94, 96–7, 173, 177, 178, 179, 180, 215, 216, 219–20, 222, 234, 250, 251, 268, 271, 273, 274, 284, 297, 311, 328
Museveni, Yoweri 196
Musyoka, Kalonzo 157
Muthaiga 68, 155, 156, 175, 190, 191, 195, 202–3, 260
Muthaura, Francis 87, 93, 177, 179, 200, 219, 222, 235, 268, 297
Muthumbi, Mary 28, 91, 221, 249–50, 264–5, 335
Mutoko, Caroline 155–8, 334
Mutua, Alfred 256
Mwai, Evan 169, 171, 219
Mwakwere, Chirau Ali 202
Mwaliko, Sylvester 93
Mwangi, Dave 84, 173
Mwangi, Wangethi 248, 249
Mwenje, David 86
Mwiraria, David 72, 85, 86, 93–4, 96, 165, 173, 179, 219, 222, 223, 250, 251, 260–1, 268, 273, 274, 284, 301

Nairobi 9, 70, 91, 103, 108, 111, 114, 117, 125, 134, 142, 145, 148, 149, 150, 154, 155–6, 167, 188; bookshops 281; cosmetic makeover of 280–1, 292–3; cosmopolitan nature of 9; economy 9; population 149; slums 2, 10, 11, 41–2, 149, 151, 157, 191, 203, 290, 303, 312, 330; State House *see* State House, Nairobi; unemployment in 55–6
Naivasha, Rift Valley 295
Nakumatt supermarket chain 292, 295, 312
Nandi language 47, 49
Nation Media Group 12, 141, 191, 248, 251, 269
National Rainbow Coalition (NARC) 261, 337; Anglo Leasing contracts, involvement in 79, 84–97, 118, 165, 166, 171–2, 173, 177, 179, 215–16, 217–20, 222, 223, 229, 245, 250–3, 255–8, 268–75; 'Artur brothers' and 256; cabinet 7–8, 72–4, 80, 84–5, 241, 244, 245, 268, 273; constitution and 8, 73, 86, 241, 243, 283; corruption within 20, 22, 79–97, 118, 165, 166, 170, 171–2, 173, 177, 179, 215–16, 217–20, 222, 223, 245, 250–1, 268;

corruption, initial fight against 2, 5–6, 7, 8, 11–12, 36, 40, 67, 73, 80–1, 82, 183, 228, 268, 314; donor community, relationship with 170, 183, 210–11; economy under 184, 220, 261, 279–82, 290, 292–3, 299–300; elections 2002 2, 3, 4–8, 14, 71, 71, 82, 320; elections 2007 300; free primary education programme 210–11; High Commissioner Clay and 334; JG's relationship with 14, 66–8, 79–87, 164–6, 171–82, 215–25, 229, 233–6, 241, 242–3, 244–5, 250–4, 268–74; Mau Mau and 111; *Nation* and 247, 269; Nyayo House and 7; lavish spending of members 79–80
National Security Intelligence Service (NSIS) 80–1, 84, 90, 95, 235
Ndii, David 16, 67, 69, 117, 292, 330
Ndubi, Haroun 53
Ndung'u Commission 65
Ngumi, John 116
Ngunyi, Mutahi 162
Nigeria 11, 16, 102, 115, 326
Njonjo, Charles 178
non-governmental organisations (NGOs) 10, 79, 141, 143
Ntimama, William 114
Nyanza province 297, 298–9, 300
Nyerere, Julius 140
Nyeri, Central Province 108–9, 111–12, 118, 126, 255, 282, 289
Nyong'o, Peter Anyang' 238

Obama, Barack 8, 44
Obasanjo, Olusegun 192
Ochola, Robert 310
Odindo, Joseph 141, 222, 248, 249, 269
Odinga, Jaramogi Oginga 73, 300
Odinga, Raila 73, 215, 241, 245, 257, 279, 282, 284, 296, 297, 298, 299–300, 301, 303, 305, 306, 307, 310, 311, 315
Odoi, Frank 7
Official Secrets Act 95, 236–7, 249, 319
Okolloh, Ory 153, 154, 158, 324
Okonjo-Iweala, Ngozi 326
Opus Dei 33, 134, 135, 174
Orange Democratic Movement (ODM) 284, 290, 297, 298, 301, 302–3, 306, 308, 309, 312, 313, 314, 338
Organisation for Economic Cooperation and Development (OECD) Convention on Combating Bribery 199, 276, 328
Otieno, Gladwell 224, 324
Otieno, Stephen 300
Ouko, Robert 26, 228, 299, 321
Oxfam 160, 189, 205

Pallister, David 327
Party of National Unity (PNU) 297, 301, 306, 307, 308, 312
Pattni, Kamlesh 62–3, 172, 268, 284, 295–6, 322
Perera, Anura 85, 94, 215, 268–9
Pinto, Pio 26
Power, Samantha 331
Prunier, Gerard 52
Public Accounts Committee (PAC) 95, 237–8, 253, 254, 257, 268
Public Officer Ethics Act 66, 283

Ribadu, Nuhu 326
Rice, Xan 250
Rift Valley 47, 49, 104, 112, 114, 140, 141, 282, 295, 297, 308
Ringera, Justice Aaron 20, 66, 217–18, 221–2, 228, 245, 250, 269–71, 272, 274, 322, 327
Rwanda 9, 115, 140–1, 196, 298, 299, 308
Rweria, Erastus 283
Ryan, Professor Terry 56

Sachs, Jeffrey 192, 205, 266
St Antony's College, Oxford 31, 32, 228–33, 235, 242, 243, 248, 326
St Mary's school, Nairobi 127–9, 133, 136, 142, 175
Saitoti, George 211, 251, 268, 273, 311
SAREAT (Series for Alternative Research in East Africa Trust) 162
Sargasyan, Artur 256–60
Satchu, Aly Khan 281
Save the Children 205

Serious Fraud Office, UK (SFO) 266, 276, 328
Sheng dialect 150–2, 310
Shihemi, Henry 128, 129
Short, Claire 72, 207, 287n
Sitonik, Wilson 93
Somaia, Ketan 172
Somalia 9, 78, 277
South Africa 11, 326
Soyinka, Wole 132
Standard Media Group 255
State House, Nairobi 12, 21, 22, 27, 32, 33, 34, 35–6, 38, 39, 65–6, 69, 70, 71, 74, 80, 81, 84, 89, 92, 93, 129, 160, 164, 172, 173, 174, 177, 179, 181, 183, 184, 202, 218, 219, 220, 222, 224, 232, 235, 238, 244, 249, 260, 270, 271, 283, 300, 304, 306, 325
Strathmore University 134, 135
Sudan 9, 131, 314, 325

Tanganyika 48
Tanzania 102, 140, 141, 150, 283, 299, 314
Thiongo, Ngugi Wa 132
Trade Bank 270
Transparency International (TI) 11, 14, 15, 16, 17–18, 36, 55, 66, 79, 80, 81, 118, 141–2, 160, 162, 175, 176, 179, 197, 119–200, 224, 261, 283, 289, 320

Uganda 9, 45, 46, 48, 68, 140, 141, 150, 195–6, 283, 299, 302
United Kingdom: aid to Kenya 187, 193–204, 205–12, 225, 261–2, 267, 275, 276–7, 337; colonial rule of Kenya 8–9, 11, 45–50, 52, 60, 104–10, 115, 122, 134, 172, 184, 194, 202, 203, 286–8; High Commissioner to Kenya 71, 171, 183, 194–204, 210–15, 224, 225, 254, 259, 267, 277, 333–4; trains soldiers in Kenya 9
United Nations 314; Africa bureaux 9; awards Kenya Public Service Award 2007 285; Convention against Corruption 65; gives Kenyan Treasury more autonomy over use of aid 286; Habitat 149; Joe Githongo's work for 124; Millennium Development Goals 207, 287; oil-for-food programme 263
United States: aid to Kenya 189, 193–4, 222–3, 224, 250, 259; ambassador to Kenya 193–4, 222–3, 224, 250, 259; Federal Reserve 263; State Department 197; warships based off Kenya 9
USAID 189

Vogl, Frank 61
Volcker, Paul 263, 264, 278
von Szek, Samuel Teleki 104

Wa Kibiru, Mugo 45
Wainaina, Eric 129
Wainana, Binyavanga 1, 106, 152, 324
Wako, Amos 170, 268
Wanjigi, Jimmy 84, 87, 91, 93–4, 129, 175
Wanjui, Joe 15–16, 175, 265, 297
Wanyeki, Muthoni 324
Warah, Rasna 67, 131
Were, Hussein 58–60, 272, 273n, 327, 334
Wolf, Dr Tom 158
Wolfensohn, James 185, 262
Wolfowitz, Paul 262–3, 275–6, 277
World Bank 66, 305; Anglo Leasing scandal, reaction to 260–1; corruption, attitudes towards 141, 170, 184, 185, 199, 207–8, 223–4, 260–1, 262–3, 275–6, 277, 278, 333; Development Committee 276; DfID and 275, 276; Governance and Anti-Corruption (GAC) Framework 276; Institute 207–8; JG joins Volcker panel 263, 278; JG's resignation from Kibaki government, reaction to 223–4; Moi era, grows wary of funding during 170; 'pushing money out the door' 189; staff intimacy with African governmental staff 190–2; structural adjustment programmes 60–1; talks up Kenya's prospects 332
World Vision 189

Zaidi, Ali 138–40, 141, 236
Zenawi, Meles 192
Zhvania, Zhurab 227–8